REPROGRAPHIC SERVICES IN LIBRARIES

Organization and Administration

CHARLES G. LAHOOD, JR.
and
ROBERT C. SULLIVAN

LTP Publication no. 19

LIBRARY TECHNOLOGY PROGRAM
AMERICAN LIBRARY ASSOCIATION
CHICAGO 1975

Library of Congress Cataloging in Publication Data

LaHood, Charles George, 1922-
Reprographic services in libraries.

(LTP publication ; no. 19)
Bibliography: p.
Includes index.
1. Copying processes. 2. Photography—Library applications. 3. Microphotography. I. Sullivan, Robert C., 1927- joint author. II. Title.
III. Series: American Library Association. Library Technology Program. LTP publications ; no. 19.
Z681.L34 686.4 75-25585
ISBN 0-8389-3166-9

Printed in the United States of America

CONTENTS

PREFACE

The need for a manual or a set of guidelines to assist those faced with the responsibility for providing reprographic services in a library was identified in discussions at meetings of the Policy and Research and Executive Committees of the Reproduction of Library Materials Section (RLMS) of the Resources and Technical Services Division (RTSD) of the American Library Association (ALA) in the late 1960s. These preliminary discussions led to negotiations with ALA's Library Technology Program (LTP) and the establishment of a subcommittee of the LTP Advisory Committee to further plan and monitor the publication of such a manual. An initial outline of the contents of the proposed manual, prepared by William R. Hawken, was discussed and revised by the subcommittee in June 1971 at the ALA conference in Dallas, Texas. Negotiations with the present authors were concluded in mid-1972 and the initial draft was completed in late 1973.

It was recognized at the outset of the planning for this publication that a full-scale technical manual was neither practical nor necessary, particularly in view of the existence of LTP's *Copying Methods Manual.* The variables existing in libraries, depending upon size, location, clientele, holdings, specialization, mission, policies, and local practices and conditions, and so forth, make it impossible to provide more than general guidelines or a discussion of significant considerations to be taken into account in establishing or administering a library reprographic service. It was agreed that the manual should be written primarily for those libraries attempting to establish reprographic services departments for the first time, and specifically for the personnel responsible for organizing, managing, or administering these departments. The intent, therefore, is to place the primary emphasis on the planning, organization, and administration of library reprographic services, rather than on technical processes or laboratory procedures and to provide general guidelines and policy considerations rather than detailed operating instructions. Because the authors' experience is with a reprographic service in a large research library, the guidelines may seem to emphasize this point of view. However, through experience over a period of many years with reprographic services in libraries of various sizes throughout the world, and particularly with the advice of the LTP advisory subcommittee, a conscious effort has been made to address as large as possible an audience of librarians and administrators no matter how small or large their present responsibility for providing reprographic service.

1

INTRODUCTION

Scholarly photocopying of existing library materials is done primarily for the convenience of the user. Whether a photocopy is requested to avoid the task of transcribing a text in hand or to avoid travel time and expense, or is acquired for addition to a library's collections, the factor of convenience is present. In a narrower sense, however, the term "convenience copying" refers to self-service copying on one of the multitude of photocopiers available on today's market. The present widespread availability of self-service copiers all but eliminates the researcher's expenditure of valuable time previously required for tedious and error-prone notetaking. In a wider sense, the convenience of the user requires availability of a vast number of documents and of rare out-of-print works. Any consideration of space versus the limits to be placed on use suggests that microforms be obtained.

The installation and supervision of convenience-copying facilities, or simply the contractual arrangement for the installation of such facilities in the library, frequently comprises the extent of the reprographic services offered or, in fact, needed. The introduction of self-service copying is at times a mixed blessing for the library, since, among other things, the patron's photocopying a book may create a strain on fragile pages and bindings, thus inflicting irreparable damage. The librarian must trust the client not to abuse the book in the process of producing a photoreproduction. Unfortunately some readers are insensitive to the needs of subsequent users of a book and are content with satisfying their own immediate need regardless of the damage that might be done to a book in the process. Librarians have an obligation to keep materials which would suffer from such treatment, such as a rare book collection, under special custody where stricter standards of user conduct can be monitored.

The machines themselves may be subject to unnecessary wear and tear by careless users. For the installation in a library of equipment to photocopy readily available materials, prudence dictates a location which offers at least a semblance of review by a responsible staff member. The "laissez-faire" attitude of some librarians toward the manner in which self-service copiers are operated by the library's clientele should be converted into one of vigilant concern for the well-being of the library's resources.

The photoreproduction of archival manuscripts and rare materials was and is a singularly apt application for photocopying inasmuch as handwritten or typewritten transcriptions separate the scholar from the original document in a substantive way. The employment of the photographic process reveals to the user the form and outline of the original. For this advantage, as well as for reasons of economy, the acquisition of a microform or hard-copy reproduction constitutes the most practical means of acquiring unique or out-of-print resources for the library. To the extent that such unique collections of data are copied by photographic processes and thus proliferated, the security of the information is assured. For this reason, librarians find the photoreproduction processes eminently useful as a security measure for the protection of rare and unique texts. To further this end, it may be advisable in many instances to substitute a service positive microfilm for the original in reference use, while reserving the master negative microfilm for the production of additional positive prints. The production of microforms as a form of insurance against the loss of public catalogs, shelflists, and similar records may also be advisable.

Photocopying for preservation purposes constitutes a most effective application of this technology for libraries and ultimately for the patrons of the library. The use of the term "preservation" in this context may be somewhat misleading since the handling of library materials for microfilming or photographing may accelerate the distintegration of a book containing brittle paper. (However, this may in some cases be more than offset by the lesser subsequent use of the original.) This type of "preservation," it should be emphasized, refers to the preservation of the content only.

Bibliographers and bibliophiles, whose interest includes the nature of the binding, the character of the paper and print, as well as other factors related to the physical nature of the book, are understandably less than enthusiastic about preservation photographing or microfilming. A comprise it is, but for a book in an advanced state of embrittlement, the options at the present time are limited. Deacidification and lamination are one option, albeit relatively expensive. The W. J. Barrow

Research Laboratory, Richmond, Virginia, has conducted extensive research in this area and has issued numerous reports. One of the best reports on the Barrow Laboratory effort as well as a history of permanent-durable paper was written by Verner W. Clapp.[1] Gordon Williams recommended cold storage for such materials: that they either be held in "suspension" for some future form of chemical restorative (which is not impossible and is actively being researched) or be microfilmed as specific titles are requested.[2] In the meanwhile larger libraries are actively purchasing such microform copies as are available, or are producing their own, either directly or through contractors, or both. The preservation programs of the National Library of Medicine, the New York Public Library, and the Library of Congress illustrate this combination of approaches to the preservation problem.

The use of microfilm has been heralded (and rightly so) as a powerful tool for reducing space requirements. Figures indicating space reductions of 90 to 95 percent are frequently cited and are generally correct. In making these estimates there is an unfortunate tendency to overlook the greater space required for the microfilm-viewing equipment and the adequate space that must be provided for the user, as compared with the reading stations needed for conventional book material. Space savings do, however, represent a substantial extra dividend over and above the other factors that argue in favor of introducing microforms into libraries.[3]

Another reason for creating microform photoreproductions is the capacity for large-scale information transfer. Examples of this are the National Technical Information Service (NTIS) program for the distribution on microfiche of government produced or sponsored scientific and technical report literature, and the Office of Education's (OE) Educational Resources Information Center (ERIC) program which distributes microfiche copies of the reports of educational research projects sponsored by OE.

Finally, photocopies in microform are used extensively for the republication of unique collections of data, out-of-print files, collections of books listed in a bibliography, or previously printed reformulated or specially indexed collections. Only a few library reprographic services actively participate in such micropublishing ventures since, in fact, most of the micropublications are produced on speculation by commercial concerns interested in marketing an edition large enough to recover preparation costs as well as a reasonable return on their investment.

2

EARLY LIBRARY USE OF REPROGRAPHIC TECHNOLOGY

"Reprography" is a relatively new term in the English language. Although it has been variously defined, perhaps the most useful definition has been attributed to Allen B. Veaner: a class of processes whose purpose is to replicate by optical or photomechanical means previously created graphic or coded messages. Reprography in the strict sense includes the processes that have been traditionally referred to as photocopying or photoduplication, as well as microphotography and multilith, multigraph, mimeograph, and various forms of offset duplicating. In the practical sense, in a library setting the full range of services encompassed by the term "reprographic" is seldom encountered. However, the broad term "reprography" has been purposely used here since it can apply to the widest possible combination of documentary reproduction services that may be grouped together in an organizational unit of a library. As used throughout this volume, the term "reprography" will refer in the main to the photographic processes of photography, microphotography, and photocopying.

Historically, libraries were not affected by the gradual evolution of the various reprographic processes until the beginning of the twentieth century. At first only the larger libraries were able to take advantage of the new processes. Even then the development of the related technologies was relatively slow until the impetus provided to technology in general by the research and development before and during World War II.

Approximately one hundred years earlier, the technological advances which were ultimately to have a significant effect on the collections of libraries, and consequently on the services provided to library patrons, began. In 1839, Louis Daguerre publicly introduced his method for producing photographic images on iodized silver plates developed with mercury vapor. Although Daguerre was not the first to produce a photographic image, his was the first photographic process to show fine detail and at the same time offer a relatively stable image. In that same year John Benjamin Dancer produced the first microphotographs with Daguerreotype plates at a reduction ratio of 160:1. Thus, the invention of microphotography occurred almost as early as the development of photography. In 1852, Dancer made a transparent microphotograph, the forerunner of the modern microfilm.

William Henry Fox Talbot is credited by many as the inventor of photography as we know it today. In 1844–46, Talbot published *The Pencil of Nature* in six serial parts, one of the first published books illustrated with photography. All of the twenty-four plates in *The Pencil of Nature* were direct photographic prints pasted into the book. Reproduction processes, such as the half-tone printing plate and similar methods for reproducing photographic illustrations, were not developed until a later date.

The full impact of photographic and microphotographic copying capability was as yet many decades away, but the possibilities of practical application of microphotographs were noted in 1853 by Sir John Herschel. Herschel foresaw too the publication of concentrated microscopic editions of works of reference such as maps, atlases, and logarithmic tables, as well as the concentration for pocket use of private notes and manuscripts.[1]

Herschel foresaw the advantages possible by reducing traditionally bulky and difficult to handle published materials to more convenient and manipulable size to facilitate handling by researchers. Instead of struggling with awkward-sized volumes of maps or heavy tomes of arithmetical or statistical tables, Herschel envisioned readers consulting significantly compacted filmed versions through a microscopic lens. The first use of microphotography on a comparatively large scale was for the delivery of messages by carrier pigeon during the siege of Paris in 1870.[2]

Modern photographic techniques began to flourish after 1871 with the discovery of the "dry plate" and its silver-bromide gelatin emulsion. The first celluloid-base photographic film was produced in 1887, and two years later Thomas A. Edison established 35mm as the first standard film gauge for nitrocellulose film. By the end of the nineteenth century some libraries had established photographic laboratories for full-size copying.[3] In 1906, Robert Goldschmidt and Paul Otlet first proposed the use of 75-by-125mm microfiche at the Congress International de la Documentation Photographique de Marseille.[4] Practical use of the suggestion, however, did not occur until much later.

Amandus Johnson, director of the American Swedish Historical Museum in Philadelphia, a pioneer in the use of photographic techniques for the acquisition of research materials, wrote:

> In 1905 I had a large number of photographs taken in the Royal Archives of Stockholm . . . . The tremendous advantage of photographic facsimiles over mere handwritten copies was evident and I gave the subject considerable thought, how I best could make my own photographs without too great expense. . . . About 1907 or 1908 I hit upon the idea of taking two of the usual legal size documents on one plate, then deciphering the plate with a strong reading glass. . . . In the Fall of 1908 I purchased a small German camera, 4.5 x 6cm (about 1½ x 2 inches), and had two small leather cases made to hold twenty-four metal plate holders each. In other words, I could now take forty-eight pages of manuscript without reloading and at a low price, as the 4.5 x 6cm plates were cheaper. . . . Either in 1910 or 1911 (my diaries and notes were destroyed in a fire in 1918) I hit upon a practical way of using unperforated films for photographing manuscripts.[5]

Thus Johnson had introduced "high reduction" photography to serve the practical needs of scholars for the acquisition of research materials from a distant depository.

The Library of Congress installed its first Photostat copier in 1912, and most university and large public libraries had them by the 1920s.[6] In 1924, Otto Barnack's Leica camera became available. It was this camera which made it possible for the scholar to take his own microphotographs, which in turn stimulated microfilm activities. Although relatively expensive, the Leica camera was lightweight and compact and had a high-quality lens; it was a precise and reliable mechanism that could be readily mastered by the layman. The film exposure area was 24 by 36mm. This camera was utilized in early experimental microcopying of newspapers in the Hoover Institution on War, Revolution, and Peace, in Palo Alto (Stanford) as early as 1926. It was also used by Lodewyk Bendikson at the Huntington Library at San Marino, California, at about the same time.[7] Although the Leica camera was in all essential respects technically suitable for microfilming research materials in the custody of libraries and archives and was much used by scholars during the 1920s and early 1930s for this purpose, it did not readily lend itself to the higher production requirements of a large-scale copying program because of its limited film capacity (5 feet of film) and the necessity to advance the film manually.

In 1937 the introduction of the Photorecord camera, with its automatic film advance, light control, and shutter release, and its 100-foot film capacity immediately made possible a substantial increase in production capabilities.

At about the same time the Leica camera was being utilized by scholars, commercial applications of microphotography were being developed. George P. McCarthy invented the Recordak camera, which was first utilized for copying bank checks on 16mm microfilm. The Recordak camera employed a rotating drum which conveyed the material being copied through a lens system to be recorded on microfilm moving in synchronization with the document. Introduced on May 1, 1928, in New York, the camera was later used to microfilm newspapers. From this beginning the modern rotary camera, widely used in microfilming today, was developed.

Developments in the use of photoreproduction by government libraries and archives for the period beginning in 1928 have been thoroughly documented by Robert D. Stevens.[8] As early as 1905, Herbert Putnam, the Librarian of Congress, evidenced an interest in securing for the Library of Congress copies of archival records in European repositories pertaining to the early history of the United States. In 1914 the Library of Congress began to secure hand transcriptions of materials in the British Museum, as well as in various repositories in France, and by 1919 documents in various Mexican collections were also being copied by hand. The copying of materials in overseas depositories by photographic methods for the Library of Congress was inaugurated on November 19, 1927, by the utilization of a Photostat machine. In 1928, Samuel F. Bemis, in charge of the Library of Congress European operations for copying historical documents relating to America, known as "Project A," installed a Lemare microfilm camera in the National Archives and a second camera at the Ministry of Foreign Affairs. The camera, developed by a Paris optician, utilized 35mm unperforated film in lengths of 16.5 feet—enough film for 100 exposures. The camera microfilm was used as a working intermediate since an enlarged positive photographic print was routinely produced from each frame.[9]

By the mid-1930s the techniques of microphotography had spread throughout the scholarly community and demands increased for libraries to establish facilities for making photoreproductions (microfilm, Photostats, or photographs) of scholarly materials available. The New York Public and the Yale University libraries had established photocopying services available to the public by the mid-1930s; the Library of Congress, Brown University Library, and the libraries of the University of Chicago followed suit before the end of the decade.[10]

Throughout the 1930s and '40s photocopying was restricted to the technician utilizing dark rooms and "wet" chemical processes. The end of World War II and the decade of the 1950s, however, marked the beginning of a revolution in photocopying which did away with the necessity for these methods. In 1950, the Thermo-Fax, the diffusion-transfer-reversal (DTR), and the Xerography (electrostatic) processes, as detailed by Hawken,[11] were first introduced in the United States. The ensuing competitive battle for control of the officer-copier field led to the availability of a plethora of photocopiers, most of them utilizing the basic patents covering the electrostatic method of photocopying. These electrostatic photocopying machines were designed to utilize either plain or zinc-oxide-coated paper. The former process employs a drum coated with a photoconductive substance capable of holding a static charge, while the latter uses a zinc-oxide coating on the paper to hold the static charge. The photocopying field, because of its sheer size (now estimated to exceed $1 billion a year), continues to attract new manufacturers employing variant or new processing techniques, and the competition in the field is intense.

3

REPROGRAPHIC PROCESSES IN LIBRARIES

Concurrent with the development and emergence of reprographic technology, libraries grew in number and size, as well as in the diversity of their collections and patrons. Because of the variety of forms of material in library collections,[1] as well as the equally varied demands by faculty, students, micropublishers, reprinters, book publishers, and other patrons, all of whom have their own specific needs for photoreproductions of these materials, it is not surprising that a wide range of reprographic services are now available in libraries throughout the United States.

To effectively plan, organize, and administer these services, librarians must be familiar with such processes and equipment as conventional silver-halide black and white, as well as color photography, Photostat process, photocopiers, microforms, (including 35 and 16mm roll film and microfiche), enlargement prints (including those produced on microform-reading machines and by laboratory equipment), and computer-output microfilm (COM).

These more common reprographic processes are reviewed only briefly in this chapter. More detailed and lengthy explanations are available elsewhere. Appropriate references to more technical and complete explanations of these processes are listed at the end of the last chapter. William R. Hawken's *Copying Methods Manual* remains the most comprehensive and useful discussion of the basic processes utilized in libraries.[2] A fairly brief, but excellent, discussion of micrographics is contained in the National Microfilm Association's *Introduction to Micrographics*.[3]

In discussing the major processes we assume that an existing document, picture, map, or microform from a library collection is to be used for reproducing a single copy in full, or approximately full, size or in moderately reduced size.

SILVER HALIDE PROCESSES

Conventional Photography. Black and white, as well as color photography, commonly used by amateur and professional photographer alike, is used for documentary reproduction in a variety of negative sizes ranging from 35mm panchromatic black and white or color film; 2¼ by 2¼ inches, up to 8 by 10 inches and even larger sizes. From the black-and-white negatives are produced fine-quality continuous-tone paper reproductions of portraits, illustrations, and paintings, as well as transparencies for projection. From color film, either of the reversal type (such as Ektachrome) or color negative, excellent renditions of original illustrations, paintings, and maps are faithfully reproduced. These fine-quality photographic prints or transparencies are in demand for use in exhibits, for classroom projection, and for a wide variety of publications. Special effects utilizing conventional photographic processes are useful particularly in dealing with problem originals. Infra red, ultra violet, or other filtering processes may be used to reproduce in legible form otherwise illegible or badly faded materials.

Facsimile Reproduction or Photostat. Also employing a silver-halide technology improved upon over a period of more than half a century, facsimile reproduction was utilized extensively in earlier years; it remains useful today for reproducing documents in either negative or positive form on paper without the need of a film intermediate.[4] The process employs a prism mechanism in conjunction with a conventional camera lens system for producing a paper reproduction of an original in a single step. In past years the process was extensively employed for copying virtually all types of library materials: books, manuscripts, music, maps, and newspapers. The equipment is versatile, with models designed to reproduce in a single exposure a full-size newspaper page measuring approximately 18 by 24 inches. The camera also allows for enlargement or reduction, in a single step, of up to 50 percent of the linear dimensions of the original documents, depending on the model used. The reproduction of successive generations of copies thus makes possible the almost unlimited enlargement of a portion of a page of text. This particular characteristic of the camera is well suited for producing large-size print from relatively small textual matter for the use of visually handicapped readers.

Originally the silver-halide-coated paper employed in this process permitted direct paper-to-paper reproduction of a document, but the copy produced had a reversed polarity from the original. Most originals consisted of black or dark text on a white background. In the first reproduction from such text, the polarity of the reproduction, while right reading (thanks to the intervention of the prism), was reversed;

that is, the original black or dark text appeared as white and the white background appeared as black. Thus, if a positive copy was desired, the procedure required a two-step process of first producing the negative copy, awaiting its processing (most cameras automatically feed the exposed paper to a continuous developing, fixing, washing, and drying procedure), and then repeating the process; the second time a positive reproduction was produced from the negative copy just prepared. In more recent times, reversal, or direct positive, paper, combined with a modified processing technique, has become available. (The paper is more expensive than the nonreversed paper.) When a positive silver-halide copy is required, even with the additional cost of reversal paper and modified processing, savings in the total cost may be offered to the customer who would otherwise be required to pay the cost of the negative. In practice, however, most users do not object to receiving a negative silver-halide reproduction. Indeed some prefer it, and the decision whether to utilize the direct positive process or the less expensive negative-from-original-document process depends largely on customer demand.

While other reprographic developments have crowded out this process for much of library photocopying, Boone's survey indicated that in 1964 some 18 of 45 reporting libraries still employed this process.[5] During the intervening years advances in technology, changing demands, and economics have undoubtedly diminished the use of this process. For the library clientele requiring full-size reproductions of maps, newspaper pages, and musical scores, as well as enlargement or reduction from original material, this process, owing to the high quality of the silver-halide print, remains a useful tool for the library reprographic service.

PHOTOCOPIERS

Most of the photocopiers, or office copying machines, available on today's market employ one of two electrophotographic processes which are based on the same principles, but which differ in the method of producing a copy, as well as in the materials used. These processes, termed electrostatic and Electrofax, represent a radical departure from the silver-halide technology. Rather than employing chemical processes, these photocopiers are based on electrical and physical phenomena. The Electrofax process employs a zinc-oxide-coated paper, and either a dry developing process or a toner suspended in a fast-drying liquid hydrocarbon. In the electrostatic process, the image is formed on a coated drum belt or plate and is transferred to ordinary uncoated paper.

At the present time a virtual bonanza of self-servicing, coin-operated electrophotographic photocopiers are available; most models utilize the Electrofax process employing the liquid hydrocarbon process.[6] The variety of brands, prices, and special features of the various models undoubtedly represents one of the more vexing problem areas for the manager of a library reprographic service. With the demand for the installation of self-service coin-operated service in most libraries being virtually irresistible, the problem is to determine first whether to choose the electrostatic or Electrofax process and then to decide which brand and model machine to acquire. While many brands of photocopiers are well constructed and generally produce adequate copies for reader reference use, important factors to consider include machine design (particularly the ability to handle bound volumes as well as loose sheets), the cost and availability of maintenance services, and the necessity, or lack of necessity, for trained operators.

Where adequate photocopying staff is not available, a concession arrangement with a reliable vendor offers a solution. Under this arrangement a contractual agreement should be drawn up to protect the interests of the library and its patrons, and also to make clear the contractor's obligations and rights. The library should evaluate the contract in terms of cost of staff time needed, and the level of subsidy or cash return (if any) to the library resulting from the vendor's profit on the installation. The ultimate charge to the library user is affected by such arrangements and may weigh differently with each library. Offering the service at the least possible charge to the user—foregoing a cash return to the library from the vendor and even providing free staff service—seems to be the rule, on the assumption that low copying costs decrease the instances of theft and mutilation.

MICROFORMS

35mm Roll Film. The historically standardized microform format for libraries continues to be 35mm roll microfilm. The reasons for adopting the 35mm-size film were, and continue to be, compelling. These include the availability of equipment (camera, processor, and printers for contact film duplicates); the maximum size of the image area (1.25 by 1.75 inches) on most planetary cameras (which allows the satisfactory reproduction of large format material such as newspapers, folios, and some maps); and flexibility (which enables the producer to copy various sizes and formats of research materials at the lower reduction ratios of 1:5 to 1:22). This flexibility assures the microreproduction of a great variety of materials on a standard size film, large enough to be easily handled by the user, at reduction ratios compatible with the characteristics of the material copied. These characteristics include such factors as type size, text orientation, fold-outs, color, and condition of paper, among others. William Hawken provides an excellent explanation of the various factors affecting the characteristics of copies.[7] These factors affect all photoreproduction processes but are especially critical for microreproduction and subsequent generations of same-size or enlarged reproductions.

As long as libraries use newspapers in microform, the 35mm roll-film format will remain the "norm" for libraries. Serial files other than newspapers, and, indeed, monographs, lend themselves to this format, especially when single or relatively few microform copies are needed. Another important factor is the almost universal library ownership of reading equipment designed for use with this 35mm film at the reduction ratios indicated above, that is 1:5 to 1:22.

16mm Roll Film. To a lesser extent, 16mm microfilm is also being utilized for reproducing library materials. The reduction ratios that must be employed with 16mm film are limited to those ranging between approximately 1:18 and 1:24 for documents measuring 8½ by 11 inches. These reduction ratios assume that the document and type size and/or typeface used to print the document are acceptable for filming at these higher ratios. The use of 16mm film requires a uniformity of document size which applies to a relatively small portion of library materials. The adaptability of some 35mm cameras and reading machines to the use of 16mm microfilm provides a basis for the use of 16mm film for those files when the reduction factors are favorable. Whether the use of 16mm roll film at from 1:18 to 1:24, in lieu of 35mm film at 1:14, provides sufficient cost benefits to warrant the intermittent loading inconvenience of using the 16mm-size film on 35mm reading equipment depends on the volume of usage and the need for reader printer capability. 16mm film is particularly suitable for reproducing 3-by-5-inch card catalogs.

Microfiche. Essentially a micropublishing technique, a microfiche should be produced in at least 25 distribution copies to be economical. The production of microfiche requires either an expensive step-and-repeat camera, capable of producing unitized fiche masters on 105-by-148mm film, or the production of roll film, either 35mm or 16mm, which must then be machine-slit to proper width and "stripped up" on a frame to create the master fiche. The National Microfilm Association's *Introduction to Micrographics* presents a lucid explanation of this process.[8] Although strips of 35mm or 16mm film can be produced for microfiche on modified planetary cameras, they may not be entirely satisfactory due to the erratic spacing between images and the resultant difficulty in producing justified lines of images conforming to the ANSI specification for microfiche.[9]

ENLARGEMENT PRINTS

The library users' continuing preference for hard copy (a full-size paper photocopy) for personal use, and the demand by the publishing industry for full-size photoreproduction from a micro-image, imposes on the library reprographic services department the requirement of installing one or more techniques for producing eye-readable hard copy from the micro-image. The more traditional process for producing enlargement prints requires the use of darkroom techniques employing an enlarger and silver-halide print paper, conventionally processed in developer, stop bath, fixer, and wash. This process, used less and less because of the inconvenience and expense, nonetheless is technically capable of producing high-quality reproductions, particularly of textual materials; and in some instances the copy made by this process can result in a satisfactory reproduction not possible from other processes.

Obviously the expense of purchasing the enlargement and processing equipment needed for this type of reproduction service can only be justified where reasonable demand for quality enlargement prints from existing microforms warrants. The process is generally compatible with the darkroom technology used for producing conventional continuous-tone photographic prints. A special enlarger is required, as well as photographic papers and chemical solutions suitable to the higher contrast of microfilm. More frequently, however, enlargement prints from microforms are produced outside the darkroom environment in self-service, coin-operated reader printers or in electrostatic enlargers, such as the Xerox Micro-printer, or Electrofax enlargers.

Those factors of diversity of input of library materials mentioned earlier in this chapter, which serve to complicate and proliferate the types of microforms, and the lack of standardization of reduction ratios, image orientation, and so forth, also complicate the process of producing enlargement prints from the array of multi-formatted microforms that may be found in the library's collections.

A British-made enlarger-printer does offer more flexibility than most, owing to its infinitely variable enlargement range extending from 8 to 24 times enlargement from microfilm or microfiche. This enlarger, however, cannot be recommended as self-service equipment since it requires the services of a trained operator.

Unfortunately for libraries, most reader-printer machines available today were originally developed for business or engineering operations where input to the system is under strict control. This equipment was initially designed for the production of business forms, parts, catalogs, or engineering drawings. The screen sizes, enlargement factors, paper characteristics, and, in fact, all controllable factors are made compatible within the entire system in order to assure a cost-effective operation for these types of materials. The dependence of libraries on such equipment has imposed the requirement to improvise and compromise in order to make available to the library user many advantages of microphotographic technology.

In considering the possible introduction of reader-printers into the library, the best course is to evaluate, from a systems-analysis viewpoint, which available type of equipment would most effectively utilize the microforms available in the library and best satisfy the user's needs. Some of the burden of such an evaluation procedure can be eased by consulting the *Library Technology Reports,* published by the American Library Association, which present detailed technical evaluations of the capabilities of a wide variety of available reader-printers.

It is also possible to purchase high-speed continuous-process enlarger-printers capable of producing enlargement prints at high speed from either roll film or microfiche. For continuous paper enlargements, from either 16 or 35mm roll film, the Xerox Copyflo is available. This unit, in the $90,000 purchase price range, is too expensive for all but those libraries with a heavy demand for this kind of service. However, when the microfilm input for this copier has been produced to optimum film density, correct reduction ratio, and image orientation, the output is of excellent quality. The possibility of utilizing this equipment for reproducing full books, journal articles, manuscript collections, and catalog cards makes it a truly useful tool for a reprographic service in a large library. In fact, several of the larger library reprographic services departments have purchased or leased one or more Copyflo copiers. Commercial Copyflo service is available from service bureaus in many areas of the country.[10] Cost studies by at least two large libraries indicate that the reproduction of books, journal articles, and so forth, by first microfilming them and subsequently reprinting them full size on a Copyflo, is less expensive, provides better-quality reproductions, and causes less wear and tear on the collections than by reproducing the same material on other electrostatic or on Electrofax copiers. These studies, of course, assume monthly volume sufficient to assure economical utilization of the equipment under minimum-charge lease arrangements.

COMPUTER-OUTPUT MICROFILM

The advent of computer output microfilm (COM) equipment has generated excitement in the microfilm and computer industries as well as among librarians. A full discussion of this development is given in Avedon's book on the subject.[11] Early forecasts of demand for this equipment appear to have been overly optimistic since applications did not develop as quickly as equipment manufacturers had hoped. This equipment, however, is now beginning to be utilized more widely by libraries and deserves continuing study. COM is finding its way into an increasing number of large libraries and is being employed for an expanding variety of applications, such as recording purchase records, process control files, and bibliographic catalogs. The purchase price or rental fee of COM equipment, as of the Copyflo process, can be justified only for large-scale demand. At the present time the best approach for a library reprographic service considering COM applications seems to be to make use of an available commercial service bureau. In some instances the combined demands of a university campus may warrant purchase or rental of a COM camera and COM film processors. These film processors are more frequently operated by the computer center personnel. Only when duplicate film copies are needed, or photographic technical advice is required, may the library reprographic staff be called upon. Technical evaluation of COM output by the library microfilm expert frequently assists the computer staff in improving the quality of the microfilm processor. Since the micrographic technology is generally not well understood by computer staff personnel, the preferable alternative is for the reprography staff to assume responsibility for processing the COM-generated microfilm.

4

SERVICES FOR THE SMALL LIBRARY

A small library, with a comparatively limited demand for reprographic services, which desires to install in-house services will find that the lease or purchase of one or two photocopying machines will adequately meet the requirements. If the library also has microfilm in its collections, some means of producing enlarged paper or hard copies will probably suffice. The moment a library reaches the decision to acquire as much as a single photocopier to satisfy the demands of its clientele, it assumes a variety of responsibilities relative to the client, to the material to be photocopied, and to the library staff.

The first decision required by the librarian is the selection of a photocopying machine from among the many brands and models available. Selection of a copier should be based primarily on the nature of the materials in the library's collections that are to be copied and the relative predominance of the various formats of these materials in the library's collections. If bound books and periodicals predominate, one type of copier may be required; if loose sheets, manuscripts, or cards predominate, another may be more suitable. If there is a combination of types, with no one type predominant, but with an appreciable percentage of the collection consisting of bound volumes, the copier design best suited to book copying should be selected. Other factors to consider include the design of the equipment in terms of handling bound volumes, the time required for an exposure, the size of the photocopies that can be produced, and the quality of photoreproduction possible from the different types of library materials. If reproduction of illustrations is a primary consideration, care must be taken to determine that the copier selected will produce satisfactory copy.

For bound material the first consideration is whether the pages of the volumes can be reproduced. Some photocopiers accept only loose sheets, and others accept only small pamphlets or single issues of a periodical; a photocopier so designed should not be considered for general library use. Another consideration is that the machine design not damage a bound volume or otherwise harm any item in the library's collections. If the library collection contains many older, fragile books, particular consideration should be given to design features which require minimum pressure on the spine of a bound volume. A good general discussion of the problems involved in copying from bound volumes is contained in William Hawken's *Photocopying from Bound Volumes.*[1] This volume, published by LTP in 1962, is still the best general discussion on the subject.

The size of the photoreproduction available from each brand and model of photocopier deserves special consideration. Many copiers offer the user a choice of copy size of either 8½ by 11 inches or 8½ by 14 inches. Thus both standard letter size as well as legal size documents can be fully reproduced interchangeably. If it is anticipated that the library's clientele will have no need for the legal size reproduction, this option need not be considered when selecting the copier; or, if an otherwise satisfactory model also provides this option, it may be modified to limit copying to the 8½ by 11-inch size. It is obvious that the option to use 8½-by-14-inch paper when only an 8½-by-11-inch copy is needed can result in a waste of 3 inches of paper, a substantial loss and expense if continually repeated.

Related to paper size is the question of whether to use paper in rolls or pre-cut sheets. The roll-stock machine can afford greater feeding reliability, particularly in areas where climatic conditions cause frequent changes in temperature and humidity. These conditions can adversely affect the feeding characteristics of some types of cut sheets unless the atmospheric conditions in the library are rigidly controlled.

Finally, the paper capacity of a photocopier is particularly important. The copier having a capacity of only 200 to 400 cut sheets may represent a service problem, particularly for evening and weekend periods when no operator is present to renew the paper supply. Many roll-fed copiers have the capacity of making up to 88 8½-by-11-inch copies; others are available with a capacity of up to 1600 feet of paper in roll form. Obviously, the greater the paper capacity, the less time it will be necessary to spend on reloading. The descriptive literature for each brand and model should, therefore, be examined carefully to determine paper capacity.

In general the electrostatic and Electrofax copiers available on today's market are technologically capable of producing satisfactory, legible reproductions of average quality originals. However, for a library installation just any copier may not prove satisfactory, because most libraries contain large quantities of material which are below average in tonal range or readability. Any copier being considered for purchase should first be tested, and the test copies should be carefully selected from the library's own collections. Test charts prepared by manufacturers are produced in colors and tone most compatible with the characteristics of their own copying devices. When available, the American Standard photocopier test chart now being developed by ANSI Committee PH-5 Subcommittee 3 should offer accurate and impartial results.

No matter how excellent the quality of a photocopy produced on a newly installed copying machine, it should be borne in mind that quality of copy can quickly deteriorate unless the machine is properly maintained and serviced according to the manufacturer's instructions. Unless properly maintained, even the best of copiers can produce poor-quality reproductions. Every machine requires regular and knowledgeable attention. This point cannot be overemphasized.

Even with only a single photocopier in the library, the personnel factor is extremely important. Neglect of the photocopier inevitably results in poor image quality and invites justifiable criticism of the library management. The on-duty member of the library staff should be a "key" operator, properly trained to operate and service the photocopier. Should the key operator not be able to place the photocopier back in service if something goes wrong, he should be given the responsibility for seeing that it is properly serviced by the manufacturer, the concession owner, or the lessor. The quality of the response to such service calls should be recorded so that these facts are available when time for renewal of the lease or concession is at hand. If the level of service warrants, the lease or concession arrangement may be cancelled as set forth in the contract. The key operator in fact must not wait for a complaint on the function of the photocopier but should make a daily check on the copier, performing routine cleaning and reloading tasks as required.

In addition to these duties, a library staff member must be assigned responsibility for ordering and obtaining the required supplies. The discovery that the last roll of paper is already in the machine and a long weekend is ahead during which the supply cannot be renewed can result in frustration both for staff and user.

If the machine or machines acquired are coin-operated, the necessity to make cash refunds in case of machine malfunction and the need for coins to operate the copier are continuing problems for the staff. If volume of use warrants, the installation of appropriate change machines can save staff time, and at the same time assist in expanding the use of the services. Some arrangement for refunding coins in case of machine malfunction is necessary. Those disbursing the coins from their personal reserves or from a small advance fund created for this purpose should itemize the disbursements and require the signature of the receiver. The volume of cash refunds may determine the extent of record keeping necessary.

Administrative or staff use of a coin-operated photocopier may be possible by a by-pass key whch overrides the coin mechanism. However, each exposure made in this way is metered by the machine. In such arrangements the holder of the key is responsible for usage in excess of cash receipts. (Copies made by the operator as part of regular maintenance service should be accounted for separately.) The holder of the key has a responsibility to exercise strict control over the key for authorized use only. When several different library accounts are to be charged for use of the copier, a separate use record is warranted. A log or individual card filled out by each user should account for machine meter reading. Special metering devices are also available for controlling non-coin users.

The library with only a single photocopier occasionally encounters the problem of priority of use. When a coin-operated, self-service copier is also used by the staff for answering mail requests or other purposes, long-run copying should be limited to avoid queuing. Staff use, when possible, should be made during those times when the library is closed to the public or during periods of little public utilization. A notice to library users should be posted regarding maximum copying time allowed should others be waiting to use the copier.

It is a fact that once the public and staff have access to a photocopier, the demand for copies soars, leading almost inevitably to the need for additional copiers. If the demand is for increased copying capability rather than for diversity of photocopying, the logical step is to add additional copiers of the same type at strategic service points. The identical brand and model should be considered, assuming, of course, that the first copier is satisfactory. The reason for this is to take advantage of quantity purchases of identical supplies (generally affording lower unit prices), and of improved servicing which a number of identical photocopiers commands. There is also the advantage of minimizing confusion for key operators and other staff members who might otherwise be faced with a bewildering variety of machines to operate and service, which can often lead to operator errors.

Copying requirements, however, can diversify through a program change or for some other reason, necessitating consideration of a more suitable photocopier or of a new configuration of copiers better able to meet copying requirements. For instance, a requirement for the reproduction of catalog cards in quantity may justify the installation of a special copier. Since the structure of fees on many photocopiers does not take the multiple or duplicating factor into account, a copier designed to meet the need for frequent reproduction of a limited number of multiple copies of the same text may also be desirable. Introduction of a photocopier permitting this type of metering can result in lower charges and, at the same time, more flexibility in the library's photocopying requirements, with consequent savings in labor and queuing time. The selection of a copier designed for either single copying, or for making multiple copies, is not an easy decision to make since the rates of each manufacturer are apparently designed to discourage valid comparison. For relatively high-volume multiple copying, equipment other than a photocopier should be considered.

A library having only a single photocopier may find itself with all the problems of the largest photocopy service. Assuming that the library services interlibrary loan requests or mail orders, there arises the need for handling cash receipts, for establishing deposit accounts, or for maintaining a record of transactions with libraries in a cooperative system. In some instances even the library with the single copier may have heavy mail-order demand from other institutions or from commercial enterprises, particularly if its collections contain rare or unique materials. The use of a deposit account system (see chapter 8) reduces paper work by avoiding billing or payment in advance, and, as an added benefit, it generally improves the time required for handling an order.

In addition to photocopying service the library owning a microfilm collection is also faced with the need to provide its users with the means of securing enlargement prints or reader-printer copies. While high-quality enlargement prints are produced on special enlarger equipment made for darkroom use by a trained technician, the small or medium-size library in all likelihood would not have sufficient demand to support this type of service. For these libraries the solution is to employ one of several microform reader-printers modified for coin operation, or, if not coin-operated, monitored by a staff member who is charged with collecting the copying fee.

The reader-printers currently on the market and suitable for use in libraries are capable of reproducing good to poor photocopies. Here again close attention to the servicing of the equipment and experience in producing satisfactory photocopies depend on the librarian or attendant in charge, who may also have to serve as the key operator.

The selection of the reader-printer, as with the photocopier, should take into account the nature of the library collections to be copied. In reproducing from microforms, the compatibility of the reduction ratio used in producing the microforms with the enlargement ratio of the reader-printer is essential. The capability of the reader-printer to accommodate the various sizes and formats of microfilm in the library's collection is an equally essential consideration.

Library Technology Reports has reviewed most of the currently available reader-printers suitable for library use. These reports and any other available evaluations should be consulted before committing the library to a particular brand or model of reader-printer. Librarians should be aware that salesmen or commercial representatives for reader-printer manufacturers are not fully informed about the extensive and varied requirements of libraries. It is unfortunate that the commercial and the library applications of microforms and microform equipment are frequently incompatible.

The reprographic services discussed in this chapter are those suitable for the smaller libraries. For all but a few of these libraries, the photocopiers or reader-printers required are self-service, which may include the capability of producing transparencies for use on overhead projectors and copies on different colors, weight, and sizes of paper (including the possibility of some enlargement or reduction of copy from the original).

Rarely would a small library establish a facility for producing microfilm; and perhaps less likely that conventional continuous-tone photographic darkroom processes would be required. The relatively infrequent requirements for such service in a small library cannot ordinarily justify the outlay required for such equipment and for the necessary technical expertise. For these libraries the reprographic services of larger institutions must be utilized or perhaps the services of campus or local photographers. In many instances established and reliable microfilm service bureaus may also be available.

5

SERVICES FOR THE MEDIUM-SIZED LIBRARY

For the medium-sized library with a higher and more varied demand for reprographic services, the logical expansion beyond the self-service copiers and reader-printers is into the area of microfilm. The ability to produce 35mm roll microfilm provides the library with a powerful tool to extend its services both to the library users and to the management needs of the library itself. Perhaps first in importance is the ability of microfilm to disseminate information at a relatively low cost, either in the form of republication of existing material or of original publication. It also provides a means for preserving worn or fragile material from the collections, and, finally, it provides a substitute for the interlibrary loan of printed or manuscript materials. As a tool for library management, the microfilm camera can be used to produce, as an insurance copy, a microfilm of unique files, catalogs, or other records essential to the operation of the library.

There are three types of cameras used for producing microfilm: the rotary camera, the step-and-repeat camera, and the planetary camera. The rotary camera is designed to film material in the form of loose sheets at high speeds. It is restricted in the size of documents it will accommodate, and it cannot be used to film books or other bound materials. The step-and-repeat camera is a specialized camera used to create microfiche by exposing a series of images on an area of film in a predetermined pattern of rows and columns. The third type of microfilm camera, and the most versatile single piece of photographic equipment available for microfilming, is the planetary camera. When used in conjunction with commercial or other off-site film processing, a planetary camera, with its built-in variable film advance, can substantially expand the range of reprographic services available in a library. It can photograph documents in a wide variety of sizes and formats, including bound volumes, at variable reduction ratios. It can produce single exposures, or roll film or, with the aid of certain ancillary equipment, it can be used to produce microfiche. With a planetary camera the array of products and services can be extended to include not only the production of master microforms, from which working intermediates or distribution copies may be printed, but also the production of negatives from which continuous Xerox Copyflo enlargements can be made. It can also produce negative film as an intermediate for conventional darkroom photographic enlargement prints or reader-printer reproductions, and as an intermediate for the reproduction of card catalogs. It can also be used for audiovisual (slide) production and, finally, for the production of color microfilm slides.

While the planetary microfilm camera is an extremely useful tool, it is at the same time a relatively complex instrument, best able to produce satisfactory results if operated by a supervisor or technician who possesses skill and technical competence. The proper operation of a planetary camera in a library requires a good understanding of microphotography, particularly the copying of textual material found in library collections. This understanding must extend to the characteristics of the types of microfilm currently available, including their sensitometric specifications and their potential for reproducing, in fine resolution, a wide density range even on one document. With precision and reliability the skilled operator can overcome many of the difficulties inherent in microfilming a file of documents. Lacking this essential knowledge, an operator will produce microfilm of, at least, uneven quality. Even a skilled operator's film will occasionally fail to meet desired standards and have to be rejected; an amateur microphotographer's level of rejected film may reach intolerable levels. Professionally produced microfilm is a credit to a library; and to the reader who has to use it, the quality of the microfilm is all too visible.

Unfortunately, the task of recruiting a skilled microphotographer, especially one who also understands the peculiar requirements of library collections, is not easy. The more normal procedure in a library microfilm service is to hire a willing worker and either train him on the job or, lacking a supervisor with the required knowledge, arrange for the necessary training in a commercial facility. Training courses in microphotography are now being offered in some urban areas under the auspices of technical schools and various governmental agencies. The sources for training should be investigated before the expense of individual instruction by a commercial facility is incurred.

Minimum qualifications desirable for a trainee microphotographer, in addition to a high school education and the general clerical aptitude to read and understand instructions, including detailed technical and bibliographical specifications, include manual dexterity, physical stamina (to stand all day if necessary), mechanical aptitude, and the ability to concentrate. Certainly an interest in photography, the desire for a career in a library, particularly as a reprographic technician, and evidence of a methodical and organized nature would be desirable attributes as well.

To qualify as a skilled technician, a camera operator must evidence a concern for adherence to standards of quality and the ability to produce microfilm within a narrow range of tolerances in film resolution, in background density, in the positioning of documents in the image area, in illumination, and in other factors directly bearing on the quality of the finished product. These and many other technical factors are described in the specifications compiled by the Library of Congress Photoduplication Service and published by the Government Printing Office.[1] These publications, used in combination with Allen Veaner's *Handbook*,[2] comprise an essential informational source for a neophyte camera operator as well as for a library reprographic service manager.

Exposing the microfilm in the camera, while important, is but the first step; the film must next be processed, then inspected, and, finally, corrected if necessary. The library with only one or two microfilm cameras should not normally attempt to develop, fix, wash, and dry the microfilm in-house, but rather should make use of the processing facilities of a reliable commercial laboratory—preferably located reasonably close to the library. Mailing or shipping exposed but undeveloped microfilm over long distances can be done, but it is at times both risky and inconvenient. The hazards include subjection of the film to excessive heat, moisture, x-rays, and light. The primary reason for a library of this size to send its microfilm out for processing, however, is economy. Purchasing and operating a continuous microfilm processor which will produce consistently fine images capable of meeting American National Standards Institute standards for permanent record microfilm is a relatively expensive undertaking. The critical point for considering in-house processing must be carefully considered and will vary for each library. Only a demand for processing several hundred feet of microfilm a day would begin to justify in-house microfilm processing. Furthermore, it is unrealistic to expect 100 percent success in automatic film processing. It is an inescapable fact that continuous microfilm processors, despite careful attention, can and do occasionally break down and, for this and other reasons, damage or destroy the usefulness of camera negative microfilm. (This is one reason why it may be impossible for a library to commit itself to a specific deadline for completing a work assignment for a large microfilm order.)

Inspection of the processed microfilm should be completed as soon as possible. This is essential to identify possible camera or operator failure as evidenced on the processed film, and to enable the prompt reshelving or refiling of the documents that have been filmed. Many critical factors can be determined by visual inspection: legibility, proper positioning of image, completeness, inclusion of targets, and, finally, overall cleanliness of the film—freedom from dirt, chemical, and water spots. However, some technical factors, such as adequate washing, proper background density, and so forth, can only be checked by accepted technical procedures. The proper washing of film to result in a maximum acceptable level of residual thiosulfate, for example, can only be determined definitively by the methylene blue test (ANSI PH 4.8-1971) or by the silver-stain test also described in the same standard. The acceptable range of background density must be checked on a correctly calibrated densitometer.

A thorough discussion of the production of camera-negative microfilm is contained in the National Microfilm Association's *Inspection and Quality Control of First Generation Silver Halide Microfilm.*[3] An excellent general discussion of the necessity of and the techniques for the inspection of processed microfilm is presented in Allen Veaner's *The Evaluation of Micropublications.*[4]

If a library does decide to process its own microfilm, it is essential that the reprographic service department with in-house film processing, as well as coin-operated photocopiers and microfilm cameras, be managed by a full-time, technically qualified employee, plus additional trained personnel, either full or part time. The level of demand, as well as the total number of hours the reprographic service is open to its public or is committed to respond to service calls, will determine the total staffing needs.

A reprographic service, not unlike other technically oriented disciplines, has need for expert support for the maintenance and repair of its sophisticated equipment. Some preventative maintenance and even some minor repairs should be undertaken by the in-house staff. While this effort may entail an investment in specialized hand tools and test equipment, such expenditures pay dividends when the timely repair by staff materially assists the completion of promised photoreproductions. However, one must distinguish between those pieces of equipment or tools which are essential to operating the reprographic service and those which are required infrequently or are merely useful. The extent of purchase of supportive tools and equipment depends on the size of the operation, the length of time and distance from outside repair services, and—always—availability of funds.

Selection of type and brand of equipment either for purchase or lease requires a careful analysis of requirements. Basic technical data and illustrations of available equipment in the field of micrographics are readily available in Hubbard Ballou's *Guide to Microreproduction Equipment.*[5] The National Microfilm Association also publishes annually a *Buyer's Guide to Microfilm Equipment, Products and Services.*[6] This *Guide* includes lists of service companies, consultants, storage equipment, and many other useful and appropriate lists.

The availability of office copiers and toners is published annually in *Office Products News.*[7] The list is partially illustrated and includes the names of manufacturers, model number, type and size of paper used, as well as pricing, rental, and lease costs, and other pertinent data.

6

SERVICES FOR THE LARGER LIBRARY

Every library should have an overall policy statement, or a statement of the library's mission or goals, which identifies the public it wishes to serve and indicates the level of service it feels obliged and able to render to satisfy the needs. This statement should include the nature and extent of reprographic services to be provided. The determination of this major policy decision will naturally depend upon many factors and must be made by each library on an individual basis, depending on the particular set of circumstances unique to that institution and its patrons. The extent of the reprographic services to be provided by a library is a serious decision for a library administrator and should be made only after careful consideration of all the factors involved.

The widespread availability of Electrofax and self-service, coin-operated photocopiers has resulted in dramatic changes to many library reprographic services. The 1973 edition of the *Directory of Library Reprographic Services*[1] indicates the general availability of photocopier services, but a comparison of the five editions of the *Directory* published since 1959 reveals the diminishing variety of such services in most larger libraries. Relatively few libraries now offer "full service" reprographic facilities as described above. This lessening of the variety of services available evidences the expanded use of photocopiers and the decreased demand for conventional wet photographic processes. The omnipresent photocopier has reduced customer orders for conventional photographic services to the point that many libraries can no longer justify the expense of providing this fuller range of services.

A full-range "in-house" reprographic service, as often required in a larger library, may be expected to offer any or all of the following products: negative and positive microfilm; silver-halide hard-copy enlargements from microfilm; continuous-tone copy negatives; continuous-tone contact and enlargement prints; Photostatic copies; black-and-white transparencies; color transparencies; black-and-white slides; color slides; and individually or continuously produced electrostatic enlargement prints from microfilm. This list does not represent the full spectrum of reprographic services and processes currently available. It does, however, include those products which are more generally required by library patrons needing a single reproduction of research material. Infrequently requested types of photoreproduction may more economically be "farmed out" in most instances to a custom photographer or photographic service bureau.

Some of the major factors to be considered by larger libraries in determining whether to expand the service beyond photocopiers, reader-printers, and microfilm to include more photographic services are the additional professional photographic personnel, space, equipment, and related facilities required by darkroom facilities.

The decision to add darkroom facilities within the library must be based on the level and nature of the demand for such services. Once some darkroom facilities are made available, one should normally expect that additional reprographic services will be demanded. Establishing darkroom facilities involves a commitment to meet not only routine requests for photographic services but, inevitably, requests for specialized and custom services as well. The availability of services that include microfilming tends to lead to a demand for black-and-white, as well as color, slides and eventually for continuous-tone photography, the production of photographic contact and enlargement prints, and possibly color transparencies or color prints. The decision to establish darkroom facilities within the library, therefore, ought to be based on the expectation of hiring one or more professional photographers. It would be unrealistic and unwise to attempt to operate these facilities without recruiting the professional expertise required to perform the complex and responsible duties involved.

Furthermore, establishing in-house darkroom facilities requires a definite commitment in terms of space, supporting utilities, and specialized equipment. Darkrooms must be specially designed and built to avoid the accidental incursion of unwanted light, either direct or reflected. Whether required for general-purpose photographic film processing, contact or enlargement photographic printing, or color film processing, they must all be furnished with special sinks, water supply, and electrical controls. A visit to a local photographic studio or other darkroom facility may serve as a useful introduction to the complexity of an in-house photographic laboratory. Plans for the installation of darkroom facilities should include, if possible, plans for future expansion. Sound investment in the acquisition of sinks, cabinets, tables, plumbing, and electrical power lines will facilitate economical and efficient future expansion if necessary.

If demand does increase and space is available, the need for additional facilities can best be met by establishing a network of single-purpose darkrooms. Special darkrooms for discrete photographic operations, such as negative film processing, contact and enlargement print production, color film processing, and so forth, are most desirable. Multipurpose darkrooms may appear to be more economical, but fixed lighting and equipment arrangements in separate rooms, dedicated to specific forms of production, are preferable if space is available. Not only will separate quarters facilitate production, but the possibility of spoilage or waste, particularly during film-developing periods, will be reduced or eliminated if separate rooms can be dedicated to specific requirements.

Temperature control of all films and paper processing activities within the darkrooms is essential. Ambient temperature control is required, not so much for the comfort of the staff but for the protection of photosensitive films and papers. Temperature is an important factor in ensuring consistent quality photography and maintaining effective quality control. A darkroom temperature range of 68 to 75 degrees Fahrenheit is recommended. The control of relative humidity in the darkrooms is also essential. It should be maintained in the 40 to 50 percent range if at all possible. Lower relative humidity gives rise to static electricity which can play havoc with cameras, printers, and processors. Static charges may process into "tree shaped" patterns on film and can result in much wasted effort and supplies.

Whereas the average reprographic service exists to produce single copies of research material, as this is what most clients need, some users require a photoreproduction facility capable of further reproduction; for example, a color transparency used to reproduce a plate for a published book. This factor is of special concern to the reprographic service since it imposes a special need to assure adequacy of quality commensurate with the purpose intended. The highest quality is required in these instances and, if it is to be met, imposes the need for custom photography and microphotography. Assumption of the responsibility for providing custom photographic services in turn imposes on the administrator special requirements for recruiting and retaining the professional photographic personnel necessary and for revising the fee schedule to adequately compensate the reprographic service for this added investment in equipment and staff.

Staffing the reprographic service department in a larger library presents a particularly difficult problem, since the availability of the desired combination of librarian and technically competent photographer-microphotographer is extremely scarce. Nor have the schools of library education been prone to offer such a combination of training, although in the past few years there has been more interest in offering seminars to professional librarians. The seminars and the few relevant courses offered in library schools do help the professional librarian gain some general knowledge in the field; they do not and cannot, however, turn out a trained photographer. In practice, then, either a trained librarian becomes interested in documentary reproduction and eventually becomes reasonably knowledgeable in the field through practical experience, even though not necessarily a practitioner, or a photographer is brought into the library photocopy service and may eventually complete training as a librarian or take some library education courses. The important point is that the photographer working in the library reprographic service should develop some knowledge of bibliographic problems, a feel for what is required in properly handling library materials, and a deep sense of service to the users of the service.

There are professionally trained photographers and professionally trained librarians. There is need for both professions, when working in the library, to learn enough about the other's profession to offer intelligent service to the client. Nonprofessional members of the reprographic staff, such as microphotographers, film editors, processing technicians, and key operators, will probably require on-the-job training since formal instruction in these areas is almost nonexistent. A basic knowledge of photography is desirable for all members of the photocopy service staff. Unfortunately, even this requirement must frequently be overlooked when recruiting. In-depth on-the-job training, combined with whatever formal training is available, over a period of time can eventually produce the desired result of a knowledgeable, efficient, and dedicated staff.

With reference to the question whether a library should set up its own in-house reprographic service or send its work outside to a commercial concern, the in-house reprographic service, developed in a businesslike manner, brings with it definite advantages: control by the library over the handling of the materials in its collections to be reproduced; establishing priorities on services to be provided to the library's patrons; and, finally, quality control of the products provided to the library's clientele.

Some years ago the handling of library materials for reprographic services was restricted to professional photographers and especially trained photographic technicians. It should be obvious, as stated previously, that self-service copier usage needs to be monitored in some way so as to minimize damage to the collections.

Consideration of whether or not to entrust the reprographic services to an in-house staff, whose first duty should be to safeguard the materials, sometimes follows an unhappy experience with a commercial service. This does not mean, of course, that a commercial service bureau may not possess the necessary skills and knowledge either to perform the work required in its own facility or to make a special trip to the library. Since all rush orders or orders with special requirements need special handling and are therefore expensive, another consideration in engaging the services of an outside photographer is the added expense of special visits to the library if the rush orders cannot be accommodated in regular trips to and from the library.

7

PROCESSING ORDERS

Every library reprographic facility should maintain a conveniently located and clearly identified public counter or office. This centralized station for receiving orders and inquiries, even in the age of self-service copiers, is vital to the reprographic service. It is the critical point of contact between the client and the service which substantially determines the public relations image of the service. The satisfactory handling of a client's request and the resulting good public relations are the aim, and care must be taken to ensure that the client's needs are properly served. Even when a client may be unreasonable or in error, the importance of properly explaining the policy of the library and of responding to a proper request with the desired product at the stated deadline cannot be overemphasized. Therefore, it is most important that the personnel staffing the public-service counter not only be courteous but also knowledgeable of the technical capabilities of the reprographic service; this staff must be able to relate the service's capabilities to the material to be copied, as well as to the client's needs. Inquiries there, as those at a public reference desk in the library, are sometimes difficult to interpret. The form in which the initial question is posed may be a far cry from the real question. Very often the patron does not know what type of photoreproduction he wants or needs. There must be intelligent and sometimes extensive dialogue between the counter personnel and the customer. Admittedly this can be an expensive procedure, requiring much tact and patience, but it is nonetheless essential if the client is to be well served. A client may be writing a book and require a photoreproduction suitable for submission to a publisher. Unless the counter clerk can elicit this special requirement, the customer may be frustrated and unhappy with the service provided. Patrons must be encouraged to volunteer as much information as possible when placing an order so that appropriate service can be provided.

A reprographic service producing photographs, slides, color transparencies, microfilm, enlargement prints, and such acceptable for normal uses should not, without specific instructions, be expected to deliver photoreproductions suitable for exhibition or publication. Ideally, therefore, a laboratory producing master negative microfilm, for instance, should know the ultimate use to be made of copies produced from that negative in order to properly determine the technical requirements for the negative itself. The casual client's real needs must be elicited, if possible, and determining them is the essential skill that must be learned by the service-counter personnel. If a customer's requirement is beyond the knowledge of the counter personnel, they should have technical expertise readily available to them, either from a counter manager or, if absolutely necessary, from the technician who will actually process the order. However, it should be pointed out that it is an important duty of counter personnel to protect the technicians from interruptions. While laboratory personnel are generally willing to give advice, it is preferable if it can be given to the counter clerk "behind the scenes" and then be relayed to the customer to minimize expensive interruptions. The risk of loss or damage to material being processed in the laboratory is surely increased by such interruption, not to mention the loss of production.

While the needs of clients should, to the greatest extent possible, be anticipated in terms of the range of services available, this is not always possible as a practical matter. The many forms of library materials, combined with the peculiar requirements of individual patrons or academic departments, assure an almost endless variety of demands. Here, too, the judgment of the public-counter personnel may be required to solve a difficult problem. A conference with the custodial department and the technician may be necessary to determine the proper approach to a specific request. If the request is for a product not offered by the reprographic service, the service has the option of accepting the client's order with the understanding that the material will be sent out to a commercial photographic laboratory or of advising the client where the required service can be secured and allowing him to make his own arrangements. There ought to be a policy to guide this type of decision since it should be determined in advance which types of materials may leave the library to be copied and which firms can be entrusted with library materials. Channeling the client's request through the library reprographic service will usually result in a more satisfactory product for the client since the client can benefit from the library's staff's evaluation of the quality of the outside laboratory's product. Of course the client should expect to pay for this kind of service, and the library should make what it considers a reasonable additional service charge for providing it.

In general, the public-counter client should be urged to deliver the material to be photocopied to the counter. The delivery of materials results in savings to the reprographic service if the service is responsible for bibliographic searching, and it may be desirable to pass this savings along to the client in the form of a reduced fee. However, the main advantage to the client is the expeditious service possible. Since the public-counter clerk may spend considerable time completing the required order forms and any other forms that may need to be inserted in the materials to be copied, it is generally preferable to establish a fee for counter service which includes this routine, as distinct from the fees for bibliographic searching and delivery of the material to the reprographic service. It should be noted that the delivery of the material to the counter by the client is not always possible since certain library materials, such as rare books and manuscripts, generally may not be removed from their custodial departments.

The counter clerk must be not only an accomplished interviewer but a patient teacher and a diplomat as well. Educating patrons to the various rules and regulations of the service in a polite and constructive manner is obviously desirable and, if performed consistently, minimizes dissatisfaction and later complaints. The counter clerk must be alert to recognize the differing levels of demands by students, faculty, the public, various library departments, and the representatives of business and industry. While a student may rank the cost and speed of service as the highest priority and be quite indifferent to the form of reproduction provided, a faculty member may regard the quality of the reproduction as his paramount need. Representatives from business or industrial firms, or other persons not familiar with library services, often need to be referred elsewhere in the library. They may not be fully aware of the library organization and of the various services available in other departments, and a photocopy may not be necessary to fill their needs. Also, it should be recognized that all requests are apt to be presented in an urgent tone. The counter clerk must be able to withstand a certain amount of pressure and be able to diplomatically explain why a particular type of order may require several days more than the customer's expected deadline. Firm but realistic promises of completed work must be given and the counter clerk should follow up to see that these promises are kept.

Orders received by the reprographic service through the mail are generally more difficult and expensive to process than those received from the counter client. When an order is submitted in person (assuming that the counter clerk who processes the order performs his duties properly), there should be little room for misunderstanding or difficulty in processing the order. However, the order received by mail lacks this added exchange of information and requires careful scrutiny and interpretation to ensure that the needs of the requestor are correctly stated on the order form. If the order form is carefully designed, it should minimize the difficulties of interpreting a mail order. However, the clerk processing the mail order must be careful in reviewing it to ensure that the form is clearly and fully completed. If the order is unclear, incomplete, or unacceptable in any aspect, it may be preferable to return it to the sender to clarify his request rather than risk filling the order improperly. While it is expensive to conduct correspondence and to process an order a second time, it may be preferable to do this than to risk supplying the wrong material which may result in an even more costly exchange of correspondence, the return of the unwanted material, and related accounting and public relations difficulties. It is important to recognize that mail orders are more difficult and expensive to process and must be screened carefully before being forwarded to the searcher or to the laboratory.

How much, if anything, the reprographic service department should charge the client for the added cost of processing a mail order depends on who is responsible for searching the library's collections to identify the material to be copied, to secure it from the shelves, and to mark it so that the technicians who actually produce the copies requested can process the order correctly and expeditiously. If the reprographic service employs a searching staff to perform these essential functions, then the overhead borne by the service is significantly increased. If the order can be referred to other organizational units of the library, to at least identify and retrieve the material, the cost to the service is markedly reduced. The decision about which unit of the library provides this vital service of identifying, locating, marking, and physically transporting the material is one that must be made by each library depending on its size, organizational structure, overall reference policy, and related considerations. Whatever decision is made, it is important to realize that the staff of the reprographic service must generally depend upon the expertise of the reference specialists who have the professional knowledge of the collections, the card catalogs, the serial record, and other bibliographic tools.

Proper control of incoming correspondence, particularly if there is a high volume of it, is essential. A basic consideration is which organizational unit in the library should assume responsibility for handling the order correspondence. Control of this correspondence may, in smaller reprographic services, merely consist of a record log listing the date of receipt and the date of final disposition; in a larger service, a fairly elaborate system of separate control cards for each order may be required. A control card system may include the assignment of an order or estimate number to each piece of correspondence, and the preparation of a 3-by-5-inch control card that includes a consecutively assigned control number, the name of the requester, the date, the customer order number (if any), and any related data helpful in identifying the request. In general, the organizational unit responsible for producing and delivering the requested photocopies should be responsible for the initial control of the correspondence, even where another office, such as a reference or custodial unit of the library, may be called on for assistance. Experience indicates that with this unified control, correspondence will be treated with more consistent and prompt consideration. An operational unit not responsible directly to the client may tend to minimize the priority of the request.

If the volume of mail orders is modest, the necessary searching duties may be absorbed by the regular stack-retrieval personnel employed to secure material for a reader. If the volume of mail orders justifies the employment of one or more full-time staff members to identify, locate, and retrieve the material requested for copying, this arrangement is preferred, since it allows the service to establish and review the priorities of processing these requests and administratively supervise the searching operation. Full-time bibliographer-searchers or research assistants employed by the reprographic service can better inform the service of the work load and provide information on the status of any order in process. Whereas a public-counter clerk has an opportunity to anticipate and resolve problems while interviewing the customer, a bibliographer-searcher processing a mail order must wait to act until he can examine the material. He must inspect the material to determine whether or not it is the correct item. The searcher must have an excellent knowledge of the library's reference tools, card catalogs, serial records, and so forth to confirm that the material requested is held by the library. He must also have a thorough knowledge of the organization of the collections to determine the location of materials recorded, and a sound knowledge of the capabilities and limitations of the photocopying equipment to be employed to determine whether the order is appropriate for processing. In other words, once the searcher processing a mail order has located the material, he must give both the order form and the material a thorough review to confirm that they are satisfactory to ensure that the client can be provided with the type of reproduction requested. Again, the public relations factor is an important consideration, for the searcher must take care to avoid forwarding an order that can result in an unsatisfactory product and in client frustration.

If the bibliographer-searchers are on the payroll of the reprographic service, their salaries will be paid from the fees collected for the photocopies provided. A happy compromise for the client, however, is for the library to absorb the cost of this essentially reference service from the payroll of the regular library staff and to have these employees under the administrative supervision of the reprographic service. This policy decision is another that has to be made individually by each library.

An important question is not only who is to perform this essential as well as time-consuming and expensive searching, and who is to pay for it, but also how extensive a search service is to be provided in response to requests for photocopies. Many inquiries may be received that are not firm orders but merely requests for quotations of the availability and cost of materials being considered for photocopying. Other inquiries may be requests for price quotations for different forms of photoreproduction being considered (microform versus photocopy, for example). While some libraries respond to all of these varieties of requests on a so-called free-of-charge basis, it is obvious that the staff required to provide this information must be paid, and that the cost of this searching and cost estimating can be very sizable. If all of the expense of providing this information is to be borne by the reprographic service, then the rates charged for actual photocopies provided must absorb the costs of this service. Because personnel costs are by far the highest of all of the expenses that a reprographic service must bear, the trend in libraries, and in all institutions providing similar service, is to pass at least some of the cost on to the customer. Many libraries charge separately for all cost quotations for photocopies. Others will only perform routine searches of a maximum of approximately twenty minutes per order without added charge. A procedure that has worked well to avoid the expense of maintaining and referring to a file of cost quotations is to return all correspondence to the patron with the request that it be returned when a firm order is submitted. Most libraries will not conduct extensive searches without collecting a minimum searching fee of approximately $5 per hour searched. Again, this policy determination must be made by each individual library.

Some libraries differentiate between requests made by students, faculty, institutional, and library patrons and those made by business or industrial concerns. Depending upon local requirements, it may be necessary to honor requests by the first with no fees or reduced fees, while exacting a higher fee from the last type of client. Since providing estimates and price quotations is a very expensive procedure, care must be taken to ensure that the time invested in processing such requests is reasonable and commensurate with the anticipated return or fee to be collected if an order is placed. For example, a searcher cannot, except in unusual instances, be allowed to invest an hour's time on a request for a quotation for photocopies of two pages of a book. If a client requests a quotation on the cost of supplying photocopies of 100 different books, it might be sufficient to estimate the cost of a representative group of five or ten items only and to return the request with the schedule of rates for those items which will allow the inquirer to project his own estimate for the remaining items.

Care must be taken also to treat general inquiries, which do not request a specific item to be copied, as reference inquiries to be referred to and processed by a library department other than the reprographic service. If the inquirer wants a selection made of the best item available to be copied, for instance, then the inquiry should be treated as a reference question to be responded to by the appropriate reference unit in the library rather than by the reprographic service. This type of determination must be made by the correspondence clerk or searcher before valuable time is invested in conducting a search.

Delivery to the technician of material needed to complete a mail order client's request is more than a problem of mere transportation. The bibliographer-searcher has a responsibility to mark the order and the material to facilitate the technician's understanding of the work to be done without the technician having to review the order form in its entirety. Since the diversity and complexity of the library's collections frequently requires the expertise of a subject or language specialist, this knowledge may be clearly beyond that likely to be possessed by a photocopying technician. It is essential that the material be clearly and fully identified with the order to expedite its processing by the technician. Even with the most careful marking of the material there will be occasions when the laboratory technician is perplexed by a specific request. The aim, of course, is to minimize such occasions. The consistent use of suitably designed markers and labels is indispensable to facilitate such communication and to speed the processing of orders.

In addition to possessing the ability to make the proper identification of the material requested and to determine the type of photoreproduction process required, the searcher must be knowledgeable about the schedule of fees, and be able to apply it intelligently to all types of requests. The stated policy of the reprographic service should be that it is the responsibility of the searcher or counter clerk to ensure that the least expensive copy consistent with their requirements is supplied to the clients. Fee schedules can be somewhat complex, particularly in a large library, and the full understanding and uniform application of the rate structure by all members of the staff is an important training and supervisory responsibility of the manager of the reprographic service.

Bibliographer-searchers face a particularly vexing problem in coping with orders with incomplete or inaccurate citations. Mail orders are frequently prepared by part-time student help or by others who either are unfamiliar with a library or are simply careless in describing the material to be copied. Although this is a problem encountered to some degree by all library searchers, it is of increased importance in a reprographic service since charges are made for the product supplied. Just as the counter or correspondence clerk may be well advised to return an order that is not clear, rather than risk supplying the wrong material, the searcher must exercise similar caution. All his ingenuity and bibliographic "sixth sense" must be mustered to solve some riddles presented by garbled or incomplete citations, but it is generally better to return the order for clarification than to risk supplying the wrong material. Repeat customers may specify that the searcher exercise his judgment if the cost of the order is less than $10, but for more expensive orders it is preferable to require him, if the order is ambiguous or unclear in any way, to return it with a request for clarification. Transposition of the letters of an author's surname, or even the incorrect first name, may be safely corrected by the searcher if the balance of the entry is accurate and provides sufficient identification. However, extensive editing of the citation by the searcher is both risky and time-consuming.

Order forms should specify that the patron provide the source of the citation, the series note if any, the library call number if possible, the full author identification, title, place, publisher, and date of publication, as well as the LC card number if applicable, the ISBN or ISSN number if known, the pagination, and any other bibliographical detail available to prevent misunderstanding. Customer use of the ALA Standard Photoduplication Order Form (see appendix D) may be required of regular users of the service. Searchers have to be cautioned to bring to their supervisor's attention any orders from a customer that repeat patterns of error or that are consistently incomplete, so that the originating office can be contacted and educated to complete the order form properly. This area of difficulty illustrates why it is preferable to have the searching personnel administratively supervised by the reprographic service so that bibliographical problems can be promptly detected and resolved, and reported and corrected, if possible.

8

FEES AND ORGANIZATIONAL CONSIDERATIONS

Now that we have traced the history of the development of reprographic technology, identified the principal processes and the settings for their use in libraries, and reviewed the processing of orders, we turn to one of the major policy questions in operating a library reprographic service: what are the major considerations in structuring a schedule of fees to be charged for the services provided? A number of factors must be taken into consideration, such as a possible subsidy and the costs of equipment, labor, materials, and overhead.

In the past many libraries, particularly large college and university libraries, have felt it necessary to emphasize their service function and to hold fees for reprographic services as low as possible. In fact most libraries, to a lesser or greater extent, have emulated this policy and have subsidized the operation of their reprographic services. Samuel Boone's survey in 1964 of library reprographic services indicated that the level of subsidization decreases roughly in proportion to the increased size of the library and the resulting greater level of demand for services.[1] The larger the library and the clientele it serves the lower the subsidy. The range of subsidy generally extends from 0 to about 75 percent. Only in one instance, reported by Boone, was the reprographic service self-sustaining to the point that service fees generated surplus funds to help support other library services less able to operate on a self-sustaining basis.[2]

Although there seems to be no fixed pattern for managing the fiscal affairs of library reprographic services, the level of fiscal support to be given by a library depends on the policy of individual libraries, the availability of funding, and, as mentioned previously, the level of demand. Almost every library to some degree acknowledges that it should provide the user with a measure of "cost free" (to him) service. The difficulties of adhering to this ideal for heavily used reprographic services have led to the need for service fees to recover more and more (and in some instances all) of the out-of-pocket expenses, including the costs of most management services. This approach is born of practical administrative necessity since the alternative of providing "free" services in response to an increasing demand would strain the budget of even the largest library. In practice, wish as we might that the reprographic services could be offered "free of charge" to the user, practical necessity deters even the most liberal library administration.

In determining a fee structure, and indirectly the level of subsidy, it is important to be aware of the true nature of the various expenses involved. Hawken provides a convenient checklist of the principal cost factors that includes the costs of equipment, direct labor, materials, and overhead.[3]

The cost of equipment in a normal commercial undertaking represents a capital investment. Capital improvements in the form of equipment exceeding a fixed amount of money—for example, $100—are intended to have a useful life expectancy of from three to ten years, during which time the equipment is used in some manner in the production of income-producing products or services. The purchase of such equipment usually involves a substantial expenditure at the time of acquisition, which is expected to return the investment over the life of the equipment. Equipment cost, then, is a substantial factor to be taken into consideration in establishing a fee structure. If library policy permits a full or partial subsidy for the acquisition of equipment, obviously this factor can be omitted or reduced in determining the fee structure. From a commercial accounting point of view an allowance for depreciation of equipment is essential. The portion of any fee collected for depreciation of equipment should be held in reserve so that funds are available at the time a replacement is required. If the operation is fully subsidized, no such cash reserve is accumulated. When equipment is to be replaced, the library must be prepared to expend the necessary cash or rely on the support of its parent institution. If the library or its parent institution operates on one-year funding, and if the accumulation of a cash reserve is not possible, the library must then request funding in its regular budget.

In computing depreciation costs, the entire cost of the equipment, including any auxiliary items, such as camera lenses and other nonexpendable items, should be computed. This total figure should then be divided over the anticipated useful life of the equipment. The expense of a piece of equipment costing $6,000, with an anticipated useful life expectancy of five years, is easily computed at $100 a month. In order to recoup production costs, this figure is then divided by the number of units expected to be produced by the equipment. If the anticipated monthly volume is 100 units, the equipment cost is computed as $1 for each unit produced. This figure may then become a cost factor in computing the per-unit fee. To this may be added the cost of equipment maintenance, supplies, and repairs. A service contract can simplify determination of this cost factor since the known cost of the service contract can be included in cost computation.

It is important to determine the useful life of a new piece of equipment, particularly one involving new technology. Professional accountants can offer useful suggestions on which to base a judgment. The tendency among librarians, unaccustomed to strict business practices, and perhaps sympathetic to the idea of charging the lowest possible fee to the needy scholar and student, may be to overestimate the useful life of equipment. On the other hand, an overly cautious approach suggests the utilization of a shorter anticipated useful life than is realistic. A substantial error in either direction may tend to work a hardship on either the reprographic department's budget, if too long a period is used for depreciation, or on the user, if too short a useful life span is employed. The former may result in an unrealistically low unit charge, while the latter approach may yield a charge that is too high.

Related to this discussion is the question of equipment lease or lease/purchase, as opposed to outright purchase. The obvious advice is to do that which better serves the interests of the library's reprographic service (and, of course, the library itself). What are the "best interests" of the service? In any given situation the following factors should be considered: availability of capital for purchase (if not available, the decision, of course, becomes much simpler); total outlay for purchase, compared to total outlay for leasing; anticipated life of the equipment; the rapidity with which the equipment under consideration might become obsolete due to technological improvements; and, finally, the costs of preventative maintenance and repairs.

The availability of required capital may be the determining factor in this situation. The purchase of a needed piece of equipment and supporting attachments would obviously be detrimental to the cash position of the reprographic service if the result would be that insufficient funds would be available for staff salaries or supplies. In this situation the possibility of leasing the equipment seems more desirable, provided that anticipated monthly revenue from use of the equipment will at least satisfy the minimum lease charges, the value of supplies used, and all other out-of-pocket costs. One drawback of lease arrangements, involving monthly lease charges, is the possibility of a widely fluctuating seasonal demand which could result in monthly charges to the library in excess of receipts. Such a condition should be considered in determining the unit charge. The charge should be fixed in such a way that the receipts for the higher-volume billing periods will cover the expenses of the lower-volume periods. Lease costs frequently are geared to a return of the lessor's capital investment during a three-year period. Any equipment with a useful life expectancy of three years or less, or any equipment which might be rendered obsolete owing to technological advances during a similar time period, ought to be strongly considered for a lease arrangement. If the equipment is expected to have a useful, up-to-date life of more than three years, it should probably be purchased.

A second major factor to be considered in determining fees is the cost of direct labor. This is generally the most expensive component in the array of costs relating to the products of a reprographic service; this is particularly the case for that large portion of library-related photocopying where special handling and low volume of production are the norm.

In determining direct labor costs two factors should be considered: the cost of the total labor package and the level of productivity. The first factor is frequently computed by multiplying the annual employee wage by 125 percent. The added 25 percent should include the cost of such fringe benefits as vacation, sick leave, health insurance, contribution to Social Security or other annuity plans, life insurance, and so forth. In reality the 25 percent factor, once considered the norm, may in fact be outdated and may well have to be increased to fit the circumstances of the individual library.

The second direct labor factor is related to the efficiency of each production worker. Even assuming a "full-time" effort on the part of a production worker, it is clear that the standard eight-hour work day is in actual practice something less than that. Tardy arrival, extended rest and lunch periods, time taken for telephone calls, and other reasonable absences from a work station, all constitute significant time factors in the work day. For a nominal eight-hour work day, some administrators compute productive time at seven hours. Even this figure may be too optimistic in an easy-going work environment. To realistically compute direct labor costs, one must honestly assess the total work situation. To ignore these facts is to flirt with either fiscal insolvency or the imposition of unrealistically high fees for the library user. In other words, reasonable adherence to productivity levels should be maintained regularly by each worker in order to sustain the fee structure, which must in the final analysis be comparable with the rates of other library reprographic services.

Another factor to consider in constructing a schedule of fees for reprographic services is the cost of materials. This factor too deserves careful analysis since it requires control of costs at three points: purchase, storage, and use. Efficient purchasing practices should include taking advantage of quantity discounts that will result in tangible savings during the time the quantity purchased is used. It makes little sense to purchase small quantities of material when higher quantity purchases provide lower unit costs. On the other hand, it would be wasteful to purchase a two-year supply of photo-sensitive materials which have a shelf life of one year. Purchasing in quantities that can be conveniently and safely stored is clearly indicated. Maintaining strict inventory control can result in cost savings by signaling the need to purchase additional stock before the inventory is exhausted and also by serving notice on all staff that supplies are not available for personal appropriation.

In computing material costs, anticipated normal wastage caused by human error and by material and machine failure should not be overlooked. It is inevitable in a reprographic service to sustain a certain amount of wastage. In the production of photographic enlargement prints, operator judgment in securing the required quality of reproduction cannot be expected to be infallible. In film processing, continuous film-processing equipment may damage the film. Cameras may malfunction, resulting not only in loss of film and processing time but also the added loss of the camera operator's time. While the manager of every efficient reprographic service desires to hold losses to a minimum, a factor for waste of film and processing functions should be included in his calculations. Attempts to utilize so-called nationwide averages for such losses are of questionable value. There is no easy formula and, therefore, every reprographic service should determine its own loss record for the various processes and compare these, if they appear excessive, with results obtained at comparable installations.

A further factor to be considered in computing service fees is overhead cost, which is usually distinguished as being either direct or indirect. The computation of overhead costs provides the librarian the most likely opportunity for providing at least some subsidy for the reprographic service. It also presents the area least clearly understood by those not required to account in strict businesslike detail for all cost factors.

Direct overhead includes the cost of direct supervision, clerical costs, and direct administration. A larger organization may also include the cost of handling incoming mail or counter orders, bookkeeping, and financial reports, as well as statistics related to total productivity, cost analysis for orders, general correspondence, and so forth. Other direct overhead charges include the value of laboratory, storage, and office space; cost of communication, such as telephone, teletypewriter, telefacsimile, and postage; and normal housekeeping services such as the wrapping and labeling of outgoing mail.

Mailing costs, a considerable expense attached to the processing of mail orders, must be taken into account in constructing a schedule of fees. Postage costs alone have risen sharply in recent years. Containers to properly wrap and safeguard materials in transit also represent a sizable expense. Photoreproductions that are prepared to exacting specifications can easily be rendered useless if not carefully wrapped for maximum protection during shipment. The full cost of postage and insurance must be charged in addition to the regular fees for photocopying. The quality control exercised throughout the bibliographical and technical processing of the order must be extended to include the collating, wrapping, and mailing procedures to ensure customer satisfaction. Special handling or shipping instructions stated on the order must be observed and the expenses incurred in following these instructions passed on to the patron. Particular care must be taken when the shipping address differs from that of the ordering office so that the photocopies are delivered to the specified address. Mailing and shipping containers should be purchased in standardized sizes and in quantity to minimize expenses. It is important that good business practice be observed in all of these operations if service is to be effective.

Indirect overhead costs which may also be included in the fee structure comprise two major factors: first, the bibliographic identification and the physical locating and retrieval of materials requested for replication, and second, the quotation of costs to the client for the service requested.

The principal elements of direct cost can be determined only after a detailed analysis of all relevant factors has been conducted. The details of a cost accounting procedure should be formulated in cooperation with a qualified accountant, developing a methodology for establishing accurate cost figures. The methodology must include a projection of the volume of annual demand for each type of service, the productivity levels attainable by a given staff, and the inclusion of all other direct and indirect costs involved in operating the service. Once reasonably accurate cost figures are developed, the library administrator will have enough facts to determine the level of any subsidy that might be necessary and, consequently, the fee schedule. The need for realistic cost accounting therefore cannot be overemphasized, for such accounting directly affects the fiscal viability of the reprographic service. Ignoring any significant cost factor, large or small, may result in a cash shortage and possible insolvency of the reprographic service.

The preferred arrangement for reimbursing the reprographic service for services provided is some form of advance payment. This may seem to be an unduly harsh policy but it is a realistic requirement that flows from the nature of the service provided. A businesslike operation should be conducted in a businesslike manner. The product generated for each request is tailored to the needs of each individual requestor. Each order is filled on a custom basis. The photocopy is produced individually each time, rather than being produced in quantity in advance and being removed from a stock of ready-made items. The possibility of efficiently and economically storing and servicing an inventory of photocopies is practically nonexistent. The problems of the maintenance of any inventory, and of the space required for it, for other than the very few "best sellers," make it impractical to attempt to anticipate demand by stocking any given items. For the same reasons returns for credit cannot be accepted. The cost of storing and searching such returns against subsequent orders is prohibitive. Therefore, a "cash and carry" policy should prevail.

A further incentive for a policy of advance payment is the high clerical cost of billing and maintaining an accounts-receivable file. The average charge made for a completed order is relatively low and simply does not justify the expense of trying to collect payment after delivery has been made. Exceptions to the advance payment requirement may have to be made, particularly for institutions that are precluded by government regulations from making advance payments. Arrangements may be made to accept purchase orders and to bill upon delivery, perhaps with a separate invoicing charge added.

For the regular patron, particularly another institution, corporation, or governmental agency, the best fiscal arrangement may be a deposit account system. An advance deposit of funds, to be supplemented regularly, should be made to cover adequately the anticipated volume of orders with a minimum of paper work, and without the need for a prior cost estimate, thereby expediting the processing for both parties. The reprographic service can provide a statement to the customer with each transaction, or at stated intervals (at least quarterly), showing the cost of each order and the balance of funds remaining in the account. The arrangement should also include an agreement that no order exceeding an agreed-upon maximum cost will be processed so as to preclude the possibility of a request being processed which will cost the client far more than he anticipated. It would obviously come as a shock to a customer to be charged $100 for a reproduction for which he estimated a cost of $10. Another control that can be exercised is to instruct the customer to state a "do not exceed" cost on the order as a signal to the service not to proceed with the order if the actual cost will exceed the maximum cost indicated. Counter orders should be paid at the time they are placed or, at least, upon delivery. Mail orders should refer to a deposit account number or include prepayment.

Another form of prepayment is the coupon plan. This plan also offers advantages to both the customer and the service. However, its use is generally practical only where the format of materials in the collections are so similar as to permit copying units to be standardized in size and where the types of photocopying processes available are quite limited. This situation prevails where a relatively uniform collection of documents exists from which the photocopies are prepared. The National Lending Library in Boston Spa, England, utilizes such a coupon system very successfully. This government-sponsored documentation service, now part of the British Library, until recently copied mainly journal articles. A more typical library photocopying service, offering a greater variety of forms of photocopies from a more heterogeneous collection of library materials and with more varied requesting sources, will probably find a coupon scheme impractical. A recent survey by a committee of the American Library Association investigating the possibility of adopting a coupon plan in American libraries met with little enthusiasm because of the wide range of materials and services available, fees applicable, and fiscal regulations imposed by institutions as well as by local, state, and federal government agencies.

No single payment policy is practical for a library reprographic service. Again, a policy decision to fit the particular circumstances is required. Probably a mixed policy, requiring predominantly advance payment, coupled with the acceptance of limited unpaid purchase orders from select customers will prove to be practical. The handling of accounts receivable, it must be emphasized, not only adds the extra costs of billing and accounting but, given prevailing interest rates, the cost of money itself. The fiscal character of the service may be such that it could find itself unable to meet current expenses if it accumulates any quantity of outstanding invoices, even though these may be 100 percent collectable. The library reprographic service must be operated in a sound businesslike manner if it is to survive and be able to continue to provide the service for which it was created.

Departing from fiscal considerations, we turn to the question of organizational location. The most logical answer to the question of where a library reprographic service should be placed in the table or organization of the library is that it should be located where it can best serve the needs of the library and its clients and, at the same time, promote the safety and preservation of the library materials to be reproduced. In most cases the library reprographic service should be located in such a way that the equipment and operating personnel are concentrated in as few places as possible. A centralized service should promote maximum utilization of equipment and enable the library to purchase more efficient and economical equipment, such as higher production capacity, more mechanized photocopiers, cameras, and film processors. Services centrally organized with full responsibility for all library reprographic work can offer a superior range and quality of service.

The library administration responsible for the management of a reprographic service should appreciate that the proliferation of separate and scattered organizational units is a luxury that most organizations can ill afford. On at least one large university campus there are approximately 200 darkroom installations located in departments throughout the institution. The rationale behind these separate installations reportedly is that each department is "doing the job free." A close examination of the operations of this fractured organizational arrangement indicates that the quality of the work is generally poorer and more expensive than that afforded by a centralized reprographic service.[4]

Organizationally a reprographic service may be considered a reference, technical, circulation, administrative, or other service. Because in many instances it serves an interlibrary loan function, it is frequently placed under the supervision of that department. Again it must be said that this decision should be made depending upon the particular circumstances found in each library. A primary consideration, other than to centralize the service if at all possible, is to recognize that its operation affects all areas of the library and that its management should be closely coordinated with all other operations.

9

OTHER ADMINISTRATIVE, BIBLIOGRAPHICAL, AND TECHNICAL CONSIDERATIONS

A number of policy considerations, in addition to those considered earlier, should be entertained by the management of a library reprographic service. These may be grouped generally into administrative, bibliographical, and technical considerations.

The preceding chapter treated the general administrative considerations in operating a library reprographic service, such as the questions of a fiscal subsidy, accounting practice, fee structure, reimbursement procedures, organizational location, staffing, and training. Reference was also made in chapter 1 to the need to educate the staff to extend their responsibilty to conserve the library's collections through the monitoring of material being photocopied. Just as a binding priority should be given to the most-used materials, so a replacement or preservation priority should be assigned to material most in demand. However, it may be necessary to refuse to service an item for a reader during an interim period until its replacement is secured.

Restrictions on the use of photocopies are also necessary, and the librarian has a responsibility to enforce them. The photocopying of certain categories of materials, such as United States currency, bonds, postage stamps, and so forth, is illegal. A list of these categories is available from many manufacturers or lessors of photocopying machines (see also appendix B). Certain materials in a library's collections may bear published restrictions against their being copied and these also should be honored. Materials may be deposited or given to a library with the understanding that they may not be reproduced without the written consent of the owner. By accepting this material for addition to the collections, the library incurs the obligation to honor the restriction. Whatever restriction is applicable should be clearly stated on the material to ensure its observance.

Copyright restrictions must also be observed. "Fair use" is still being defined, or it has been agreed not to attempt to further define it in the deliberations under way at present in connection with the proposed revision of the United States copyright law. This term is given a latitude of interpretation, but, clearly, extensive photocopying that would deprive authors or publishers of the sales of publications still in print should not be allowed in a library. In an attempt to limit liability, some libraries require the completion of a form by the patron (see sample in appendix C) before accepting an order for photographic services of copyrighted material. Signs should be prominently placed near every photocopying machine warning patrons not to violate copyright. The importance of displaying this notice to limit the librarian's liability cannot be overemphasized. What actually constitutes a copyright infringement can probably only be determined in a court case, but nevertheless, librarians are not free to permit or tacitly condone what appears to them to be a blatant disregard of copyright. The number of copies prepared by, or for, any one customer must be restricted to a single copy and needs to be monitored to prevent reproduction of multiple copies as well as the copying of sizable portions of current materials that are presumably still for sale in the original published version since this copying would likely constitute an infringement of copyright.

The library can review counter and mail orders effectively and reject an order for current copyrighted materials not accompanied by written permission of the copyright owner, if it judges that copyright clearance is necessary; but the monitoring of self-service copies is more difficult. The latter can only be reviewed by the librarian providing the material to be copied, or by the librarian responsible for supervising the machine. One deterrent is to limit the production of the number of copies at any one time. Admittedly, this type of restriction is virtually impossible to police effectively. The librarian must be aware of the existing copyright law, follow the progress of cases on copyright that are pending in court, as well as of the pending legislation on the revision of the copyright law.

Another form of restriction that the library reprographic service may wish to employ is to reserve the right to supply a positive print in lieu of a negative. The obvious benefit possible from applying this policy is that it affords the librarian an opportunity to retain a negative photograph or microfilm of an item in the collection making it unnecessary for the original item to be copied thereafter. Subsequent requestors of photocopies can be supplied a positive print prepared from the master negative retained by the library rather than disturb the original material. This policy should only be applied on a selective basis when the material to be copied is of such value or importance, or is in sufficiently deteriorating condition, to justify the trouble and expense of the library's retention of the master negative. The library may choose to charge the original customer the full cost of both the negative and positive in such cases; or it may consider the cost of the negative an expense to be charged to library funds allocated for preservation purposes, and charge the customer only the cost of the positive print; or the customer may be charged the negative cost and receive the positive, while the library pays the positive cost and retains the negative. Libraries do have an obligation to preserve valuable or unique materials and should exercise the right to substitute a positive for a negative when required as a preservation measure.

The storage of master photographic negatives in approved storage facilities removed from the main library building is recommended for preservation purposes as a matter of policy. It is only practical to implement this recommendation where duplicate or working negatives, of course, exist since these negatives are needed by the reprographic service to fill customer orders. The ideal situation, if sufficient funds exist, is to print duplicate negatives to be utilized for printing copies and to store the master negatives separately. The master negatives then only need to be utilized to replace the printing negatives.

Processing orders for book or reprint publishers, micropublishers, or similar commercial firms may require further policy decisions. Since these orders may necessitate literally taking a book apart to ensure the preparation of the best-quality copy possible, the library may exercise several options: it may choose to exercise the right to microfilm a preservation copy of the work for its own collections, and refuse to supply any copy other than a print from the resultant master negative; it may decide to charge a fee to reimburse the library for the expense of lending the material and making any necessary repairs afterwards; it may request a free copy of the new work, plus the expenses incurred in any repair or rebinding; or it may elect some combination of those alternatives.

The Reprinting Committee of Resources and Technical Services Division and the Association of College and Research Libraries have issued policy statements on lending to reprinters and micropublishers. Librarians should consider that all materials in need of preservation cannot be economically preserved by libraries alone and that the cooperation of all parties possible, including reputable and responsible commercial firms, should be encouraged. Fees charged should be adequate to compensate libraries for expenses incurred, but librarians should keep in mind that fees serve to increase the costs of the final products ultimately purchased by libraries. Conversely, reprinters and micropublishers should keep in mind that reasonable charges by libraries should be accompanied by reasonable charges to libraries for the works produced. Therefore, it is in the interest of both libraries and commercial firms to cooperate. Libraries should cooperate only with commercial firms that have demonstrated that they are responsible in the handling of library materials entrusted to their care and take pains to produce a good quality product from both a technical and bibliographic point of view. Any borrower who abuses the privilege should be refused future cooperation. Each library must decide its own policy based on its own circumstances and experience.

Fees or royalties may be levied for photographs or other photoreproductions utilized in commercially published works. Consideration should be given to establishing these charges also, following the philosophy outlined above. The courtesy of a citation to the source of the photoreproduction ought to be quite sufficient in most cases.

Cooperation with other libraries or institutions, particularly in preservation microfilming projects, also ought to be given serious consideration. Many large microfilming projects such as those sponsored by the Center for Research Libraries, the Library of Congress, the New York Public Library, and various college and university libraries, are not possible without pooling the holdings from several libraries to complete long runs of serial or newspaper issues or similar materials and sharing the expenses involved. The ARL Foreign Newspaper Microfilming Project and the CRL-sponsored Cooperative African Microfilm Program, South Asia Microfilm Program, and South East Asia Microfilm Program projects are good examples. The cost of a master negative microfilm may be prohibitive for a single institution to bear, but if the cost can be shared by several participants the project may become feasible. As a result of such cost sharing, more preservation work can be accomplished for the benefit of many more libraries and researchers. Many serial files are in such an advanced state of deterioration that no library would consider lending them or even servicing them to a reader within the library. Microfilming such files on a cooperative basis ensures the preservation of complete, or substantially complete, sets and widespread access to otherwise relatively inaccessable files.

The proper preparation of materials for archival quality microfilming is a time-consuming and expensive process that is generally unappreciated and underestimated. While materials serviced to a reader are merely removed from the shelves and delivered "as is," or materials requested for routine photocopying are copied as they exist in the collections, this practice is inadequate for archival or preservation microfilming. A page-by-page collation of the material should be the first step, with a detailed record made of any mutilated or missing pages or issues. Bindings should be removed if possible, any missing or mutilated pages or issues replaced with borrowed replacements (or photocopies), and the file arranged in proper order with appropriate targets inserted. Detailed specifications for the archival microfilming of newspapers, monographs, and serials are available from the Library of Congress.[1] These manuals outline the detailed procedures required for the preparation of files, bibliographic and technical targets, and related processes. Adherence to these standards is important to ensure the bibliographical, as well as the technical, quality of the product. It is important for every library reprographic service to be aware of the existence of these specifications and to insist upon their being utilized when appropriate as a matter of policy.

Photographic standards covering characteristics of film base, processing and storage, and so forth are available from both the American National Standards Institute and the National Microfilm Association. Applicable standards are cited in appendix A. Librarians should not only utilize these existing standards but also welcome the opportunity to participate in the committees responsible for their drafting and formal adoption.

A library reprographic service department can and should be of valuable assistance to the library's acquisitions department in performing tests on newly acquired microforms to determine whether they meet the technical and bibliographical standards of archival quality. The same standards in force in the library's microform laboratory should be expected to be met by the microforms acquired by the library. If the reprographic service has no "in house" microform production capability, then it should assist the acquisitions department in any way it can in examining newly acquired microforms in accordance with the procedures set forth in Veaner's *Handbook.*

Adherence to photographic and bibliographic standards not only enhances the appearance or readability of the finished product but also its utility and durability as well. Standards are designed with great care to facilitate use by the library patron, the reason for the existence of libraries. If the work is worth doing it is worth doing to the best of the library's ability, and this should be the credo of every library reprographic service.

APPENDIXES

APPENDIX A

SELECTED LIST OF NATIONAL STANDARDS*

PH 1.28-1973	Specifications for Photographic Film for Archival Records, Silver-Gelatin Type
PH 1.41-1973	Specifications for Photographic Film for Archival Records, Silver-Gelatin Type, on Polyester Base
PH 1.43-1971	Practice for Storage of Processed Safety Photographic Film other than Microfilm
PH 1.45-1972	Practice for Storage of Processed Photographic Plates
PH 4.8-1971	Methylene Blue Method for Measuring Thiosulfate and Silver Densitometric Method for Measuring Residual Chemicals in Films, Plates, and Papers
PH 5.3-1967 (R 1973)	Specifications for 16mm and 35mm Silver-Gelatin Microfilm for Reel Applications
PH 5.4-1970	Practice for Storage of Processed Silver-Gelatin Microfilm
PH 5.9-1970	Specifications for Microfiches

SELECTED LIST OF NATIONAL MICROFILM ASSOCIATION INDUSTRY STANDARDS**

MS 104-1972	Recommended Practice for Inspections and Quality Control of First Generation Silver Halide Microfilm
MS 110-1974	Operational Practices Manual

*Catalog of Standards available from American National Standards Institute, 1430 Broadway, New York, N.Y. 10018

**Standards available from National Microfilm Association, Suite 1101, 8728 Colesville Road, Silver Spring, Md. 20910

APPENDIX B

TYPES OF DOCUMENTS ILLEGAL TO REPRODUCE PHOTOGRAPHICALLY

Provisions of Title 18 of the *United States Code* seem to be the basis of prohibition of duplication of the items in the following list.* Penalties of fine or imprisonment may be imposed on those guilty of making such copies.

1. Obligations or Securities of the United States Government, such as:

   Certificates of Indebtedness
   National Bank Currency
   Coupons from Bonds
   Federal Reserve Bank Notes
   Silver Certificates
   Gold Certificates
   United States Bonds
   Treasury Notes
   Federal Reserve Notes
   Fractional Notes
   Certificates of Deposit
   Paper Money

   Bonds and obligations of certain agencies of the Government such as FHA, etc.

   Bonds. (U.S. Savings Bonds may be photographed only for publicity purposes in connection with the campaign for the sale of such bonds.)

   Internal Revenue Stamps. (If it is necessary to copy a legal document on which there is a cancelled revenue stamp, this may be done provided the reproduction of the document is performed for lawful purposes.)

   Postage Stamps Cancelled or Uncancelled. (For philatelic purposes, Postage Stamps may be photographed provided the reproduction is in black and white and is less than ¾ or more than 1½ times the linear dimensions of the original.)

   Postal Money Orders.

   Bills, Checks or Drafts for Money drawn by or upon authorized officers of the United States.

   Stamps and other representatives of value, of whatever denomination, which have been or may be issued under any Act of Congress.
2. Adjusted Compensation Certificates for Veterans of the World Wars.
3. Obligations or Securities of any Foreign Government, Bank or Corporation.
4. Copyrighted material of any manner or kind without permission of the copyright owner.
5. Certificates of Citizenship or Naturalization. (Foreign Naturalization Certificates may be photographed.)
6. Passports. (Foreign passports may be photographed.)
7. Immigration Papers.
8. Draft Registration Cards.
9. Selective Service Induction Papers which bear any of the following information:

   Registrant's earnings or income
   Registrant's dependency status
   Registrant's Court Record
   Registrant's physical or mental condition
   Registrant's previous military service

   Exception: U.S. Army and Navy discharge certificates may be photographed.

*The list is not all inclusive, and no liability is assumed for its completeness or accuracy. In case of doubt, consult your attorney.

10. Badges, Identification Cards, Passes or Insignia carried by Military, Naval personnel or by members of the various Federal Departments and Bureaus, such as FBI, Treasury, etc. (unless photograph is ordered by head of such department or bureau).

    Copying the following is also prohibited in certain states: Automobile Licenses —Drivers' Licenses—Automobile Certificates of Title.

APPENDIX C

SAMPLE COPYRIGHT "FAIR USE" PLEDGE

I certify that I require the photoduplication for the purpose of private study, research, criticism or review and I undertake not to sell or further reproduce the copy supplied without permission. I further certify that I have checked and found a copy of this work is not available through normal trade channels.

Signature____

Please print name____

Address____

_____________________________________zip_______________

Company name____

Date_____________________Clerk__________________________

APPENDIX D

ALA STANDARD PHOTODUPLICATION ORDER FORM

LIBRARY PHOTODUPLICATION ORDER FORM **A**

Requesting Library
Fill in forms; send sheets A and B to Supplying library.

DEMCO LIBRARY SUPPLIES
Form No. 253
Madison, Wis.
Hamden, Conn.
Fresno, Calif.

Supplying Library
Fill in pertinent items under REPORTS; return sheet B to Requesting library.

Date of request:

Requester's Order No. . . . .

Supplier's Order No. . . . .

Call-No.

Author (or Periodical title, vol. and year)

Fold ——

Title (with author and pages for periodical articles) (incl. edition, place and date)

☐ Any edition

Verified in (or Source of reference)

Request ☐ microfilm ☐ photoprint ☐ Other

Remarks:

NOTE: This material is requested in accordance with the A. L. A. recommendations concerning the photocopying of copyrighted materials.

ORDER AUTHORIZED BY:

REPORTS:
NOT SENT BECAUSE
☐ Not owned by Library
☐ File is incomplete
☐ In use
☐ Hold Placed
☐ Request again
☐ Publication not yet received
☐ Please verify your reference — Fold
☐ Other:
☐ Suggest you request of:

Estimated Cost of Microfilm . . . . .
Photoprint . . . . .
Please pay in advance ☐
Please do not pay in advance ☐

Please send cost estimate for
☐ microfilm ☐ photoprint

Go ahead with the order if it does not exceed: $. . . . .

Special instructions:

NOTES

CHAPTER 1

1. Verner W. Clapp, "The Story of Permanent/Durable Book-Paper, 1115–1970," *Scholarly Publishing* 2:107–24; 229–45; 353–67 (Jan., Apr., July 1971).

2. Gordon R. Williams, "The Preservation of Deteriorating Books: An Examination of the Problem with Recommendations for a Solution," *Minutes,* 65th Meeting, January 24, 1965 (Washington, D.C.: Association of Research Libraries, 1965), pp. 9–44.

3. Verner W. Clapp and Robert T. Jordan, "Re-evaluation of Microfilm as a Method of Book Storage" *College and Research Libraries* 24:5–15 (Jan. 1963).

CHAPTER 2

1. J. S. K. Herschel, "New Photographic Process," *Atheneum* (July 9, 1853).

2. Prudent René Dagron, *La Poste par pigeons voyageurs, Souvenir du siege de Paris: Specimen identique d'une des pellicules de dépeches portées à Paris.* Tours-Bordeaux: Typographic Lahure, 1870–71.

3. Alfred Gunther, "Microfilm in the Library," *UNESCO Bulletin for Libraries* 16:1 (Jan.–Feb. 1962).

4. Paul Otlet and Robert Goldschmidt, "Sur une forme nouvelle du livre," *Bulletin, Institut International de Bibliographie* 12:61–69 (1907).

5. Amandus Johnson, "Some Early Experiences in Microphotography," *Journal of Documentary Reproduction* 1:9–19 (Winter 1938).

6. Lester K. Born, "History of Microform Activity," *Library Trends* 8:349 (Jan. 1960).

7. Robert D. Binkley, *Manual of Methods of Reproducing Research Materials: A Survey Made for the Joint Research Council and the American Council of Learned Societies* (Ann Arbor: Edwards, 1936), 163–65.

8. Robert D. Stevens, "The Use of Microfilm by the United States Government, 1928–1945." Ph.D dissertation, American University, 1965. (Available from University Milcrofilm, Ann Arbor, Mich., no. 65-11, 379.)

9. U.S. Library of Congress, *Report of the Librarian of Congress for the Fiscal Year Ending June 30, 1928* (Washington, D.C.: The Library, 1929), p. 228ff.

10. Herman H. Fussler, *Photographic Reproduction for Libraries* (Chicago: Univ. of Chicago Pr., 1942), p. 36.

11. William R. Hawken, *Copying Methods Manual* (Chicago: American Library Assn., 1966), pp. 113, 129, 147ff.

CHAPTER 3

1. For a concise evaluation of the problems arising from this variety of library materials, see W. R. Hawken, "Systems Instead of Standards," *Library Journal* 98:2115–25 (Sept. 15, 1973).

2. Hawken, *Copying Methods Manual,* pp. 85–206.

3. *Introduction of Micrographics* (Silver Spring, Md.: National Microfilm Assn., 1973). 28p.

4. Generally called "Photostat." Products and equipment available from Anken Chemical & Film Corp., 10 Patterson Ave., Newton, N.J. 07860; Itek Business Products, 1001 Jefferson Rd. Rochester, N.Y. 14623.

5. Samuel Moyle Boone, "Current Administrative Practices in Library Photographic Services." M.S. in Library Science thesis, University of North Carolina, 1964.

6. Hawken, *Copying Methods Manual,* pp. 157–62.

7. Ibid., pp. 1–82.

8. *Introduction to Micrographics,* p. 14.

9. American National Standards Institute, PH5.9-1970.

10. *Microfilm Source Book* (New York: Microfilm Publishing, Inc., annual).

11. Donald Avedon, *Computer Output Microfilm,* Monograph no.4, (2d ed.; Silver Spring, Md.: National Microfilm Assn., 1971).

CHAPTER 4

1. William R. Hawken, *Photocopying from Bound Volumes,* LTP Publication no.4 (Chicago: American Library Assn., 1962). 208p.

CHAPTER 5

1. U.S. Library of Congress, *Specifications for the Microfilming of Newspapers in the Library of Congress* (Washington, D.C.: Govt. Print. Off., 1972); idem., *Specifications for the Microfilming of Books and Pamphlets in the Library of Congress* (Washington, D.C.: Govt. Print. Off., 1973).

2. Allen Veaner, *The Evaluation of Micropublications: A Handbook for Librarians,* LTP Publication no.17 (Chicago: American Library Assn., 1971). 59p.

3. National Microfilm Assn., *Inspection and Quality Control of First Generation Silver Halide Microfilm,* MS 104-1972.

4. Veaner, *The Evaluation of Micropublications,* p. 25ff.

5. Hubbard W. Ballou, *Guide to Microreproduction Equipment* (5th ed.; Silver Spring, Md.: National Microfilm Assn., 1971). 794p. (Annual supplements for 1972 and 1973 also available.)

6. *Buyer's Guide to Microfilm Equipment, Products and Services.* (Silver Spring, Md.: National Microfilm Assn., Jan. 1973). 52p.

7. "A Guide to Office Copiers and Toners," *Office Product News* (Mar. 12, 1973), pp. 23–34.

CHAPTER 6

1. Joseph Z. Nitecki, ed. *Directory of Library Reprographic Services* (5th ed.; Weston, Conn.: Microform Review, 1973).

CHAPTER 8

1. Boone, "Current Administrative Practices . . .," p. 55.
2. Ibid., p. 52.
3. Hawken, *Copying Methods Manual,* p. 297ff.
4. Jim Smith, "Does a Million Dollar Photo Facility Belong on a College Campus?" *The Professional Photographer* (Apr. 1973), p. 29.

CHAPTER 9

1. *Specifications for the Microfilming of Newspapers* . . . and *Specifications for the Microfilming of Books and Pamphlets. . . .*

GLOSSARY

The definitions of the following terms have been selected from the National Microfilm Association *Glossary of Micrographics* (MS 100-1971), available from the National Microfilm Association, and are reprinted here with permission.

ARCHIVAL QUALITY, ARCHIVAL STANDARDS

The degree to which a processed print or film will retain its characteristics during a period of use and storage. The ability to resist deterioration for a lengthy, specified time. *See also* American National Standard PH4.8.

CAMERA

A photographic device, employing an optical system, used for exposing light-sensitive material.

CAMERA, PLANETARY (FLAT-BED)

A type of microfilm camera in which the document being photographed and the film remain in a stationary position during the exposure. The document is on a plane surface at time of filming.

CAMERA, ROTARY (FLOW)

A type of microfilm camera that photographs documents while they are being moved by some form of transport mechanism. The document transport mechanism is connected to a film transport mechanism, and the film moves during exposure so there is no relative movement between the film and the image of the document.

CAMERA, STEP AND REPEAT

A type of microfilm camera which can expose a series of separate images on an area of film according to a predetermined format, usually in orderly rows and columns.

CAMERA MICROFILM

First-generation microfilm; also called the master film.

COLLATE

To combine items from two or more ordered sets into one set having a specified order not necessarily the same as any of the original sets.

COM

1. Computer Output Microfilm: microfilm containing data produced by a recorder from computer-generated electrical signals.
2. Computer Output Microfilmer: a recorder which converts data from a computer into human-readable language and records it on microfilm.
3. Computer Output Microfilming: a method of converting data from a computer into human-readable language onto microfilm.

CONTACT PRINT

A print produced by exposure of the unexposed stock in immediate contact with a master or intermediate.

CONTINUOUS TONE COPY

Photographic copy which contains a varying gradation of gray densities between black and white.

COPYFLO®

A trademark of Xerox Corporation for xerographic printers and materials.

DARKROOM

A room which is used for loading and unloading and the developing of exposed photographic film or paper.

DENSITY, OPTICAL

The light-absorbing quality of a photographic image (degree of opacity of film and blackness for paper prints) usually expressed as the logarithm of the opacity. Several specific types of density values for a photograph may be expressed, but diffuse transmission density is the one of greatest use in the case of microfilm, and diffuse reflection density is generally of interest for paper prints. *See also* American National Standards PH2.17 and PH2.19.

ELECTROFAX®

A trademark of Radio Corporation of America for a xerographic (electrostatic) reproduction process capable of creating and maintaining an image on zinc oxide-coated materials.

ENLARGEMENT

A reproduction larger than the original or the intermediate.

EXPOSURE

1. The act of exposing a light-sensitive material to a light source.
2. A section of a film containing an individual image, as a roll containing six exposures.

3. The time during which a sensitive surface is exposed, as an exposure of two seconds.
4. The product of light intensity and the time during which it acts on the photosensitive materials.

HALFTONE

The reproduction of a photograph in which the gradation of tone is reproduced by various size dots and intermittent white spaces. It is produced by interposing a screen between the lens and the film.

HARD COPY

An enlarged copy usually on paper.

IMAGE

A representation of an object such as a document or other information sources produced by light rays.

IMAGE, NEGATIVE

A photographic image in which the values of light and dark of the original subject are inverted. In a negative, light objects are represented by high densities and dark objects are represented by low densities.

IMAGE, POSITIVE

A photographic image in which the values of light and dark of the original subject are represented in their natural order. In a positive, light objects are represented by low densities and dark objects are represented by high densities.

IMAGE ORIENTATION

The arrangement of objects or images with respect to the edges of the film.

LIGHT, AMBIENT

Surrounding light; the general room illumination or light level.

MASTER FILM

Any film, but generally the camera microfilm, used to produce further reproductions, as intermediates or distribution copies.

MICROFICHE

A sheet of microfilm containing multiple microimages in a grid pattern. It usually contains a title which can be read without magnification.

MICROFILM

1. A fine-grain, high-resolution film containing an image greatly reduced in size from the original.
2. The recording of microphotographs on film.
3. Raw film with characteristics as in 1 above.

MICROFORM

A generic term for any form, either film or paper, which contains microimages.

MICROPHOTOGRAPHY

The application of photographic processes to produce copy in sizes too small to be read without magnification. (Not to be confused with photomicrography.)

MICROPUBLISHING

To issue new (not previously published) or reformatted information, in multiple copy microform for sale or distribution to the public. *See also* microrepublishing.

MICROREPUBLISHING

To re-issue material previously or simultaneously published in hard-copy form in multiple-copy microform for sale or distribution to the public.

PHOTOCOPY

A photographic reproduction, excluding microcopy, generally produced by exposing the image of an original on photographic film or paper.

PHOTOCOPYING

The application of photographic processes to produce copies, excluding microcopies, generally by exposing the image of an original on photographic film or paper.

PHOTOGRAPH

1. A positive or negative picture obtained by the photographic process involving exposure of a sensitized photographic material in a camera and subsequent processing and printing operations.
2. Any image recorded on photographic sensitized material.

PHOTOMICROGRAPHY

A photograph of a magnified image (usually made through a microscope) of a small object.

PHOTOSENSITIVE

Sensitivity to light.

PHOTOSTAT®

A trademark of the Itek Corporation for paper, chemicals, and equipment used in producing document copies on photographic paper by means of a camera. The term is incorrectly applied to photocopies produced by materials and equipment of other origin.

POLARITY

A word used to indicate the change or retention of the dark to light relationship of an image, i.e., a first-generation negative to a second-generation positive indicates a polarity change, while a first-generation negative to a second-generation negative indicates the polarity is retained.

PRINTER, CONTACT

An exposing device containing a light source and a means for holding a film in close contact with the sensitized material on which the print is made.

PRINTER, PROJECTION

A photographic printer in which the negative is projected onto the print material. The image on a print made by a projection may be smaller, larger, or of equal size. It is also called an optical printer.

PRISM

A transparent body with at least two polished plane faces inclined with respect to each other, from which light is reflected or through which light is refracted. When light is refracted by a prism whose refractive index exceeds that of the surrounding medium, it is deviated or bent toward the thicker part of the prism. Prisms are often used for rotating an image.

PROCESSING

The treatment of exposed photographic material to make the latent image visible, i.e., a series of steps consisting of developing, fixing, washing, and drying.

READER-PRINTER

A machine which combines the function of a reader and an enlarger-printer.

REDUCTION

A measure of the number of times a given linear dimension of an object is reduced when photographed; expressed as 16X, 24X, etc.

REDUCTION RATIO

The ratio of the linear measurement of a document to the linear measurement of the image of the same document; expressed as 16:1, 20:1, etc.

REPROGRAPHY

The art and science of reproducing documents.

SILVER HALIDE

A compound of silver and one of the following elements known as halogens: chlorine, bromine, iodine, fluorine.

TONAL RANGE (TONAL LATITUDE)

The relative ability of a light-sensitive material to reproduce accurately the varying tones between black and white.

TONER

The material employed to develop a latent xerographic image.

XEROGRAPHY (ELECTROSTATIC)

A generic term for the formation of a latent electrostatic image by action of light on a photoconducting insulating surface. The latent image made visual by a number of methods such as applying charged pigmented powders or liquid which are attracted to the latent image. The particles either directly or by transfer may be applied and fixed to a suitable medium. *See also* Webster's Unabridged Dictionary, Third Edition.

SUGGESTED READINGS

American Library Association. Library Standards for Microfilm Committee. *Microfilm Norms.* Chicago: American Library Assn., 1966.

———. Library Technology Program. *Library Technology Reports.* Chicago: American Library Assn.

Axford, H. William. "Courses in Reprography Offered in Graduate Library Schools." *Library Resources and Technical Services* 17:246-50 (Spring 1973).

Bagg, Thomas C. "Factors Dictating Characteristics of Systems Utilizing Microforms." *NMA Journal* 3:143 (Spring 1970).

Ballou, Hubbard W., ed. *Guide to Microreproduction Equipment.* 5th ed. Silver Spring, Md.: National Microfilm Assn., 1971. 794p. First Supplement, 1971–72; Second Supplement, 1972–73.

———. "Microform Technology" in Carlos Cuadra, ed., *Annual Review of Information Science and Technology,* 8:121–61. Washington, D.C.: American Society for Information Science, 1973.

Born, Lester K. "The Literature of Microreproduction, 1950–1955." *American Documentation* 7:167–87 (July 1956).

Chapman, Ronald F. "A Beginner's Guide to Library Photoduplication. *Library Resources and Technical Services* 16:262–65 (Spring 1972).

Clapp, V. W.; Henshaw, F. H.; and Holmes, D. C. "Are Your Microfilms Deteriorating Acceptably?" *Library Journal* 80:589–95 (Mar. 15, 1955).

———, and Jordan, R. T. "Re-evaluation of Microfilm as a Method of Book Storage." *College and Research Libraries* 24:5–15 (Jan. 1963).

Francis, Frank E. "A Librarian Looks at the Microfilm Industry." *NMA Proceedings,* 15th Annual Meeting and Convention, 15:85–90. Annapolis, Md.: National Microfilm Assn., 1966.

Fussler, Herman H. *Photographic Reproduction for Libraries: A Study of Administrative Problems.* Chicago: Univ. of Chicago Pr., 1942. 218p.

Gantt, J. G. "Establishing a Library Photoduplication Department." *NMA Proceedings,* 9th Convention, 9:241–44. Annapolis, Md.: National Microform Assn., 1960.

Gosnell, Charles F. "The Viewpoint of the Librarian and Library User" in Allen Kent and Harold Lancour, eds., *Copyright: Current Viewpoints on History, Laws, Legislation,* pp. 43–49. New York: Bowker, 1972.

Guide to Microforms in Print. Englewood, Colo.: Microcard Editions. Annual.
Hawken, William R. *Copying Methods Manual.* LTP Publication no. 11. Chicago: American Library Assn., 1966. 390p.
———. "Microfilm Standardization: The Problem of Research Materials and a Proposed Solution." *NMA Journal* 2:14–27 (Fall 1968).
Heilprin, L. B. "The Economics of 'On-Demand' Library Copying." *NMA Proceedings,* 11th Convention, 11: 311–39. Annapolis, Md.: National Microform Assn., 1962.
Holmes, Donald C. "The Needs of Library Microform Users." *NMA Proceedings,* 18th Convention, 18: 256–60. Annapolis, Md.: National Microfilm Assn., 1969.
The Journal of Micrographics. Silver Spring, Md.: National Microfilm Assn. Bimonthly.
Kiersky, Loretta J. "Bibliography on Reproduction of Documentary Information," *Special Libraries* (Spring issues, 1955–66).
Kingery, R. C. "Copying Methods as Applied to Library Operations." *Library Trends* 8:409 (Jan. 1960).
Klempner, Irving M. "The Influence of Photoreproduction on Library Operations." *Library Resources and Technical Services* 7:244–53 (Summer 1963).
LaHood, Charles G. "Microform Reading Equipment—Its Place in the Library Environment" *Proceedings,* Third International Congress on Reprography. Guildford, England: IPC Science and Technology Pr., 1971.
———. "Production and Uses of Microfilm in the Library of Congress Photoduplication Service." *Special Libraries* 51:68–71 (Feb. 1960).
———. "Reproducing Maps in Libraries." *Special Libraries* 64:19, 25–28 (Jan. 1973).
———. "The Serial Microfilm Program at the Library of Congress." *Library Resources and Technical Services* 10:241–48 (Spring 1966).
———. "Use of Microforms in Libraries." *NMA Proceedings,* 19th Convention, 19: 160–61. Silver Spring, Md.: National Microfilm Assn., 1970.
Levine, Mark, "Hard Copy Print-out in Libraries." *NMA Proceedings,* 18th Convention, 18: 270–75. Annapolis, Md.: National Microfilm Assn., 1969.
Libraries and Copyright: A Summary of the Arguments for Library Photocopying. Chicago: American Library Assn., 1974, 46p.
Library Resources and Technical Services. Annual review articles, "Developments in Photoreproduction of Library Materials" (Spring issues).
Massey, Don W. "The Management and Reorganization of the Photographic Services at Alderman Library, University of Virginia." *The Journal of Micrographics* 4:1, 35–39 (Oct., Nov. 1970).
Micro News Bulletin. Silver Spring, Md.: National Microfilm Assn. Bimonthly.
Microform Review. Weston, Conn.: Microform Review, Inc. Quarterly.
Muller, R. H. "Microfilming Services of Large University and Research Libraries in the United States." *College and Research Libraries* 16:261–66 (July 1955).
———. "Policy Questions Relating to Library Photoduplication Laboratories." *Library Trends* 8:414–31 (Jan. 1960).
NMA Buyer's Guide to Microfilm Equipment, Products and Services. Silver Spring, Md.: National Microfilm Assn., 1973. 51p.
NMA Glossary of Micrographics. Silver Spring Md.: National Microfilm Assn., 1971 (MS 100-1971).
NMA Proceedings, v. 22. Silver Spring, Md.: National Microfilm Assn., 1973.

National Microfilm Association. *How to Select a Reader or Reader-Printer.* Silver Spring, Md.: National Microfilm Assn., 1974. 30p.

———. *Introduction to Micrographics.* Silver Spring, Md.: National Microfilm Assn., 1973. 28p.

Nemeyer, Carol A. *Scholarly Reprint Publishing in the United States.* New York: Bowker, 1972. 262p.

Nitecki, Joseph Z., ed. *Directory of Library Reprographic Services,* 5th ed. Weston, Conn.: Microform Review, 1973. 105p.

Orne, Jerrold. "Microforms and the Research Library." *NMA Proceedings,* 19th Convention, 19: 56–61. Silver Spring, Md.: National Microfilm Assn., 1970.

"Photoduplication in Libraries." *Library Trends* 8:3 (Jan. 1960).

Putnam, Dean H. "The Do-or-Buy Decision: In House Versus Contractual Microform Service." *NMA Journal* 3:110 (Spring 1970).

Reichmann, Felix, and Tharpe, Josephine M. *Bibliographic Control of Microforms.* Westport, Conn.: Greenwood Pr., 1972. 256p.

Rice, E. Stevens. *Fiche and Reel.* Ann Arbor, Mich.: University Microfilms, n.d. 21p.

Rogers, Rutherford D., and Weber, David C. *University Library Administration.* New York: Wilson, 1971. 454p.

Scott, Peter R., ed. *Microfilm Norms: Recommended Standards for Libraries.* Prepared by the Library Standards for Microfilm Committee of the Copying Methods Section, RTSD, ALA. Chicago: American Library Assn., 1966. 48p.

———. "Scholars and Researchers and Their Use of Microforms." *NMA Journal* 2:121 (Summer 1969).

Seidell, Atherton. "The Place of Microfilm in Library Organization." *Science* 94:114–15 (Aug. 1, 1941).

Shaw, Ralph R. *The Use of Photography for Clerical Routines.* Washington, D.C.: American Council of Learned Societies, 1953. 85p.

Shepard, Martha. "Library Services and Photocopying." *Library Resources and Technical Services* 10:331–35 (Summer 1966).

Simonton, Wesley. "Library Handling of Microforms." *NMA Proceedings,* 11th Convention, 11: 277–82. Annapolis, Md.: National Microform Assn., 1962.

Skipper, James E., ed. "Photoduplication in Libraries." *Library Trends* 8:347 (Jan. 1960).

Sophar, Gerald J. "Copyright and Information Dissemination—The Heart of the Problem." *NMA Journal* 1:12 (Fall 1967).

Spigai, Frances G. *The Invisible Medium: The State of the Art of Microform and a Guide to the Literature.* Washington, D.C.: ERIC Clearinghouse on Library and Information Science, American Society for Information Science, 1973. 31p.

Spreitzer, Francis F. "Developments in Copying, Micrographics and Graphic Communication." *Library Resources and Technical Services* (Spring 1973).

———. "Library Microform Systems—Past, Present, and Future." *NMA Proceedings,* 19th Convention, 19:161–64. Silver Spring, Md.: National Microfilm Assn., 1970.

Stevens, Robert D. "The Use of Microfilm by the United States Government, 1928–1945." Ph.D. dissertation, American University, 1965. 206p. (University Microfilm no. 65-11, 379.)

Stevens, Rolland E. "Library Experience with the Xerox 914 Copier," *Library Resources and Technical Services* 6:25–9 (Winter 1962).

Subject Guide to Microforms in Print. Englewood, Colo.: Microcard Editions. Annual.

Sullivan, Robert C. "Library Microfilm Rate Indexes." *Library Resources and Technical Services* 11:115–19 (Winter 1967).

———. "Microform Developments Related to Acquisitions." *College and Research Libraries* 34:16–28 (Jan. 1973).

———. "1972 Microfilm Rate Indexes." *Library Resources and Technical Services* 18:30–34 (Winter 1974).

———. "Report of the Photocopying Costs in Libraries Committee." *Library Resources and Technical Services* 14:279–89 (Spring 1970).

Tallman, Johanna E. "Opinion Paper: An Affirmative Statement on Copyright Debate." *American Society for Information Sciences Journal* 25:145 (May/June 1974).

Thomson, Sarah Katherine. *Interlibrary Loan Procedure Manual,* Chicago: American Library Assn., 1970. 116p.

Treyz, Joseph H. "Equipment and Methods in Catalog Card Reproduction." *Library Resources and Technical Services* 8:267–78 (Summer 1964).

———. "The Xerox Process and Its Application at Yale." *Library Resources and Technical Services* 3:223–29 (Summer 1959).

U.S. Department of Commerce. Business and Defense Services Administration. *Microforms: A Growth Industry.* Washington, D.C.: Govt. Print. Off., 1969. 18p.

U.S. Library of Congress. *National Register of Microform Masters,* 1970. Washington, D.C.: Library of Congress, 1972. 1148p.

———. *Specifications for the Microfilming of Books and Pamphlets in the Library of Congress.* Washington, D.C.: Library of Congress, 1973. 16p.

———. *Specifications for the Microfilming of Newspapers in the Library of Congress.* Washington, D.C.: Library of Congress, 1972. 17p.

———. Catalog Publication Division. *Newspapers in Microform.* 7th ed. Washington, D.C.: Library of Congress, 1973.

Veaner, Allen B. *The Evaluation of Micropublications: A Handbook for Librarians.* LTP Publication no. 17. Chicago: American Library Assn., 1971. 59p.

———. "Micropublication," in *Advances in Librarianship* 2:184. New York: Seminar Press, 1971.

———. "A Proposed Shop Manual for Library Reproduction Services." *NMA Proceedings,* 12th Convention, 12: 207–15. Annapolis, Md.: National Microform Assn., 1963.

———. "Xerox Copyflo at Harvard University Library: A Study of the Costs and the Problems." *Library Resources and Technical Services* 6:13–24 (Winter 1962).

Williams, Harry D., and Whitney, Thomas. "Xerox-914: Preparation of Multilith Masters for Catalog Cards." *Library Resources and Technical Services* 7:208–11 (Spring 1963).

INDEX